Dead Pool

The publisher gratefully acknowledges the generous support of the Ralph and Shirley Shapiro Endowment Fund in Environmental Studies of the University of California Press Foundation.

Dead Pool

LAKE POWELL, GLOBAL WARMING,
AND THE FUTURE OF WATER IN THE WEST

James Lawrence Powell

UNIVERSITY OF CALIFORNIA PRESS

BERKELEY   LOS ANGELES   LONDON

University of California Press, one of the most
distinguished university presses in the United States,
enriches lives around the world by advancing scholarship
in the humanities, social sciences, and natural sciences.
Its activities are supported by the UC Press Foundation
and by philanthropic contributions from individuals and
institutions. For more information, visit www.ucpress.edu.

University of California Press
Berkeley and Los Angeles, California

University of California Press, Ltd.
London, England

Library of Congress Cataloging-in-Publication Data

Powell, James Lawrence, 1936–
    Dead pool : Lake Powell, global warming, and the
future of water in the west / James Lawrence Powell.
        p.   cm.
    Includes bibliographical references and index.
    ISBN 978-0-520-25477-0 (cloth : alk. paper)
    1. Glen Canyon Dam (Ariz.)—History.
    2. Powell, Lake (Utah and Ariz.)—History.   I. Title.

    TC557.C62G54   2008
    363.6′10978—dc22   2008015854

Manufactured in the United States of America

17 16 15 14 13 12 11 10 09 08
10 9 8 7 6 5 4 3 2 1

When all the rivers are used, when all the creeks in the ravines, when all the brooks, when all the springs are used, when all the reservoirs along the streams are used, when all the canyon waters are taken up, when all the artesian waters are taken up, when all the wells are sunk or dug that can be dug in all this arid region, there is still not sufficient water to irrigate all this arid region.

JOHN WESLEY POWELL, 1893

We can't create water or increase the supply. We can only hold back and redistribute what there is. If rainfall is inadequate, then streams will be inadequate, lakes will be few and sometimes saline, underground water will be slow to renew itself when it has been pumped down, the air will be very dry, and surface evaporation from lakes and reservoirs will be extreme.

WALLACE STEGNER
Where the Bluebird Sings to the Lemonade Springs

Reclamation . . . cannot be indefinitely sustained. As the irrigation system approaches its maximum efficiency, as rivers get moved around with more and more thorough, consummate skill, the system begins to grow increasingly vulnerable, subject to a thousand ills that eventually bring about its decline. Despite all efforts to save the system, it breaks down here, then there, then everywhere.

DONALD WORSTER
The Wealth of Nature

CONTENTS

ILLUSTRATIONS

Overleaf Figure 1. Maps of the Colorado River basin, showing major dams and water diversion projects (Adapted from Philip L. Fradkin, *A River No More: The Colorado River and the West,* 1995, pp. x–xi)

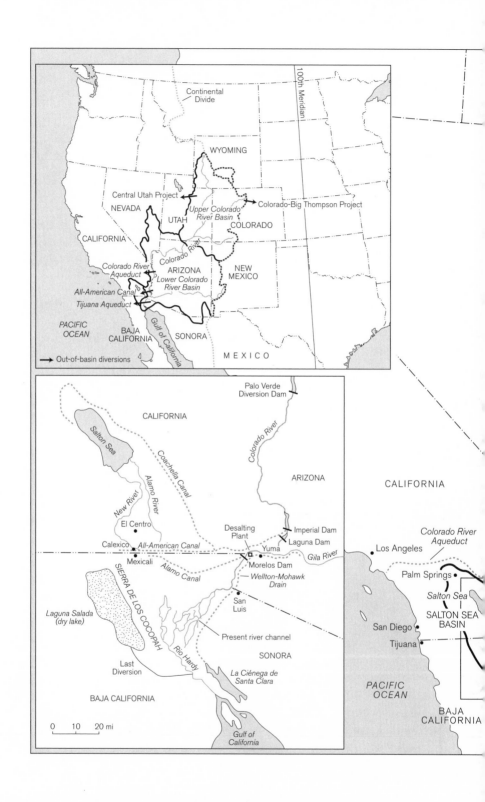

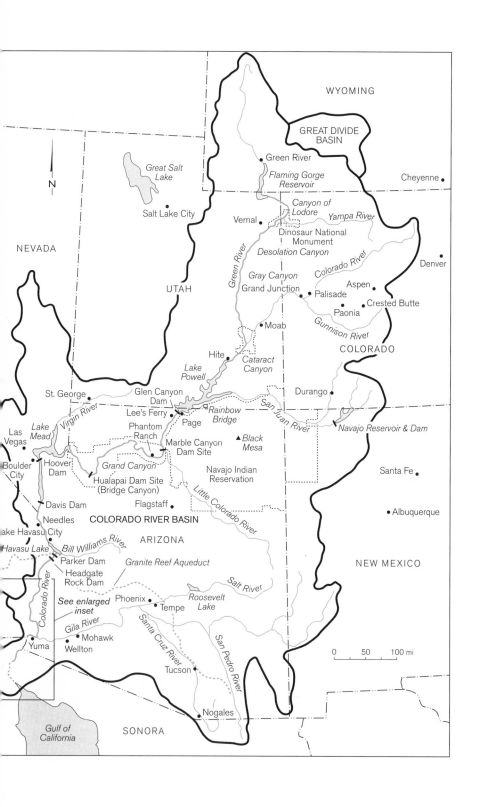

PART ONE

River of Surprise

ONE

The Dam Is Not Going to Break

BY 6 JUNE 1983, OPERATORS OF Glen Canyon Dam on the Arizona-Utah border had run out of options. High temperatures had begun to melt the late spring snowfall that blanketed the western slopes of the Rocky Mountains, sending half a million gallons of snowmelt each second rushing down the length of the Colorado River and into Lake Powell. The reservoir's two giant spillways, designed to convey high water around the dam and discharge it harmlessly below, had begun to crumble. Water entering the spillways was clear as glass, but, emerging below the dam, the water had turned red, once again earning the river the name the Spaniards had given it: El Río Colorado. Chunks of concrete, some the size of a Volkswagen, shot out with the red spillway discharge. Evidently, water under high pressure was eroding through the concrete spillway linings and into the rust-colored bedrock below, the same bedrock that held up the massive dam.

The United States Bureau of Reclamation now faced its darkest nightmare. Only a few hundred feet of soft, porous sandstone separate the spillway tunnels from the base and sides of the dam. If the spillways continued to erode, water could exploit even the tiniest opening in the weak rock, like a supersonic water-pick drilling through a loaf of bread. As the opening widened, pressure would force still more water through the opening, enlarging it, until the nine trillion gallons of water in Lake Powell drained, likely undercutting and carrying away the dam as well.

Below Glen Canyon Dam, a 580-foot tidal wave would blast through the Grand Canyon at twenty-five miles per hour, denuding its steep walls and leaving nothing alive. Three hundred miles downstream, a wall of water seventy feet high would surge over the parapet of Hoover Dam, likely causing it to collapse.[1] Each of the smaller dams below Hoover on the Colorado River's stutter-step way to Mexico—Davis, Parker, Headgate Rock, Palo Verde, Imperial, Laguna, and Morelos—would topple in turn. From Glen Canyon to the Gulf of California, the river would have destroyed each obstacle that man had placed in its path, just as it had destroyed many natural obstacles in its multi-million-year history.

Today, reservoirs on the Colorado River supply thirty million people. Glen Canyon and Hoover dams generate part of the electricity that powers a $1 trillion regional economy. Colorado River water and power sustain Las Vegas, Los Angeles, Phoenix, and Tucson, metropolises that needed only water to rise from the dusty soil of the sidewinder and the Gila monster.

When the Bureau of Reclamation completed Hoover Dam in 1935, it was the tallest dam in the world. Glen Canyon Dam, completed some thirty years later, is almost as large. Had both dams collapsed in 1983, replacing them might not have been possible. Not only would the clean-up and reconstruction costs have been enormous, neither dam could pass today's environmental reviews. Without its mega-dams, the Southwest might never have recovered.

Commencement speaker Woody Allen once advised an assembly of college students, "Graduates, as you embark on your life's journey, you will come to a fork in the road. The way to the left leads to inevitable destruction. The one to the right, to despair and misery. Choose wisely." For the managers who had to decide how to handle the high water entering Lake Powell in 1983, using the vulnerable spillways risked the destruction of Glen Canyon Dam and the other dams downstream. Keeping the spillways closed—the other fork in Allen's road—would indeed bring despair and misery, for without its spillways, Glen Canyon dam could release no more than half the water entering that June. But the reservoir was already brimful. Seen from the air, the azure lake sat in the red and buff slickrock of the Colorado Plateau like a full bowl of water teetering on the edge of a high table. If the lake could not contain all the water that entered Lake Powell, a thousand tons each second would have poured over the dam crest and destroyed the $200 million power plant at the downstream toe of the dam, ending power generation at Glen Canyon.

Tom Gamble, power operations manager at the dam, said at the time that despite the problems in the spillways, "There's no fear of jeopardy to the dam." Asked if the undeniable erosion in the spillway tunnels was a danger to the dam, Gamble replied, "We don't think so." The shaking, rumbling, and booming noises that could be heard throughout the dam were "nothing to be concerned about," he said. "The dam is not going to break."[2]

Privately the bureau told a different story: "The concern upon the June 6 report of noises from the left tunnel was for the safety of the dam and its foundation. This concern predominated throughout the spill period," said one report.[3] According to author T. J. Wolf, a bureau official warned that "any direct connection [of the reservoir through the spillway tunnels to the river downstream] could lead to erosion of the sandstone and the potential for uncontrolled release into Lake Mead."[4] For "uncontrolled release," read catastrophic flood and possible dam collapse.

The public would have been justified in taking Gamble at his word. After all, the Bureau of Reclamation is the premier dam-building agency in the world. Surely its projects provide such a margin of safety that for one to fail is unthinkable. Yet dams do fail. Only seven years before, in 1976, the bureau's Teton Dam in Idaho collapsed with fatal consequences.

No sooner had the reservoir behind the earthen Teton Dam filled than the dam fell to pieces. Eighty billion gallons of water tore downstream. The disaster forced 300,000 people to evacuate, took eleven lives, wiped entire towns from the map, and cost nearly $1 billion in property damage. According to former Bureau of Reclamation and U.S. Geological Survey geologist Luna Leopold, more than one scientist had written the bureau saying, "Look, this is wrong. You're putting that dam in a very unsafe place."[5]

Another example, though not on the bureau's watch, took place in March 1928. Los Angeles water czar William Mulholland had inspected the city's St. Francis Dam and pronounced it sound. A few hours later, the abutment turned to mud and the dam collapsed, sending a seventy-five-foot wall of water downstream. In less than one hour, 1.5 billion gallons drained. The flood killed over six hundred people and destroyed 1,200 homes. To his credit, Mulholland took true responsibility for the disaster. It cost him his career and his health and left him to die a broken man.

In the decades after the building of Hoover Dam, the Bureau of Reclamation provided pork barrel projects to western politicians and water to thirsty irrigators and cities. In the process, "BuRec" achieved God-like infallibility. One photograph from the bureau's early days captures a hard-bitten Dust Bowl couple with Model T Ford, mule team, and in the background, scrub brush as far as the eye can see. Beside them a homemade sign proudly announces, "Desert-Ranch: H. J. Mersdorf–Prop." The second line reads, "Have Faith in God and US Reclamation."[6] The sign reflected not only the couple's confidence that the bureau would take care of them, but also the presumption that the dam-builders had a biblical mandate based on Isaiah 35:1: "The wilderness and the dry land will be glad; and the desert will rejoice and blossom like a rose."

According to then-reclamation commissioner Robert Broadbent, the 1983 crisis *was* an act of God. "Sure, in retrospect we could have been releasing 30,000 to 40,000 cubic feet per second [225,000 to 300,000 gallons per second] a long time ago," Broadbent told the *Los Angeles Times*. "But how could you predict that Salt Lake City was going to have 100-degree temperatures on Memorial Day? It was the late May snowstorm and then the heat wave that caused the problem." The bureau "couldn't really have done anything differently," the commissioner explained, "except maybe save a few days, that's all."[7]

Was he right? Was the crisis unpredictable, or could the bureau have foreseen and better managed the high water? To answer, we first have to understand why the great dams and reservoirs on the Colorado exist in the first place. That knowledge will have the added benefit of helping us to judge proposals for new mega-dams, proposals that the current western water crisis is already spawning.

Before Hoover Dam, the reputation and budget of the Bureau of Reclamation had sunk so low that only a super-dam could restore them. Hoover Dam not only revivified the bureau, it made other large dams inevitable. Hoover was the first dam to rise five stories; today some 45,000 dams are as high or higher. Three decades after Hoover, Congress approved the Colorado River Storage Project (CRSP). It included another high dam at Glen Canyon and several smaller dams and irrigation projects upstream. Curiously, not a drop of water from Glen Canyon Dam runs directly to irrigators or cities. Instead the releases flow down the Grand Canyon and into Lake Mead. What then was the purpose of Glen Canyon Dam and Lake Powell? That question we will explore in depth, but one justification was that as water passed through the dam, it would spin turbines and generate electricity, whose sales would finance the entire CRSP. Glen Canyon would be a "cash-register" dam. The more water a reservoir stores, the greater the hydraulic head and force on the turbines, the faster they spin, the more power they generate, and the more cash that power earns. It was no accident that the spokesman for dam safety during the 1983 crisis was the manager of electrical power operations at Glen Canyon.

Anyone trying to manage the Colorado River first must come to terms with its unpredictable annual flow. In 1932, for example, the river carried 17.2 million acre-feet (MAF) of water. (An acre-foot equals 325,851 gallons and covers an area about the size of a football field to a depth of one foot. The choice of unit reflects the typical use of large volumes of water: to be poured onto fields.) By 1934, the river's flow had dropped by two-thirds. How much water might it carry from one year to the next? No one knew then; the best we can do now is cite the statistical probabilities. Between 1920 and 1922, the river averaged 21 MAF, flows that we now know were well above its long-term average. Unfortunately, the commissioners who met to divide the water of the Colorado River among the seven states of its basin did so in 1922, in the midst of the high flows. Misled by the wet years, they believed that the river carries at least 2 MAF more on average than today we know it does.

Lakes Powell and Mead and the reservoirs above them on the Colorado River can store a total of about 60 MAF. The bureau is supposed to reserve 5.5 MAF of capacity for incoming floodwaters. Given the river's variability, that would seem to be a minimal insurance policy. But more space means less water in the reservoirs, less hydraulic head on the turbines, and less cash generated. Therein lies the dilemma.

The mandate to the dam managers is to generate maximum electrical power while just keeping their reservoirs from overflowing. As one bureau employee acknowledged, "The system's prevailing philosophy is, keep *your* reservoir full."[8] At Lake Powell, in practice this means having the reservoir full by 1 July. The operators of Glen Canyon Dam know that several million acre-feet of snowmelt will arrive each spring and early summer, but not exactly how much or when. To monitor the mountain snowpack and forecast weather and the amount of water in western rivers, the bureau relies on the National Weather Service. But forecasting runoff is even dicier than predicting weather; runoff depends not only on the amount of rain and snow, but also on temperature, plant cover, soil moisture, slope angles, and the like.

In the winter of 1982–83, knowing that runoff predictions are inexact, logic would have dictated that the bureau should draw down Lake Powell enough to make ample room for the unknown volume of incoming water.

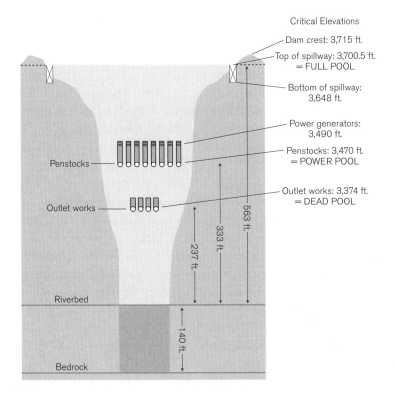

Figure 2. Schematic cross section of Glen Canyon Dam (Adapted from Glen Canyon Institute)

But those in charge of policy—officials of the Western Area Power Administration and the bureau—were judged not on how much room the reservoirs maintain but on how much electrical power the dam generates.

They had little experience with either. Lake Powell had taken from 1963 to 1980 to fill, allowing the bureau only three years of experience in managing the Colorado River system of dams and reservoirs. It was about to have to learn the hard way.

As 1983 began, the elevation of Lake Powell stood at 3,685 feet, fifteen vertical feet below the top of the spillways. The reservoir held 22 MAF and was 90 percent full. The January forecast from the National Weather Service predicted that late spring runoff would be more than 10 percent above average. To lower the risk of flooding, dam managers could have released more water to make room for the above-average snowmelt. The water would have flowed through the turbines, generated electricity, and earned cash. Instead, that January the managers let out only about half the amount they could have. Why, after receiving the forecast of higher runoff, did the bureau hold water back? What was it waiting for? The answer may have been, peak power rates.[9] Charges for electricity rise with demand. A kilowatt-hour sold in July, when air conditioners from Tucson to Los Angeles are running full blast, sells at a higher rate than one sold in January.

—

An unusually wet fall in 1982 had filled Utah's lakes and reservoirs. By February of 1983, Terry Holzworth, flood control director of Salt Lake County, began to release water from Utah Lake to make room for the incoming snowmelt. "We prepared all through the winter," said Holzworth. "We also had an extensive program to get the public ready. With the April snow, we realized the magnitude of what we were facing by the first of May—but we were way ahead by then."[10] Salt Lake City is home to one of the five major western offices of the Bureau of Reclamation. Senior officials of the bureau live there, read its newspapers, and cross its bridges. During the crisis, water from flooding City Creek was routed down State Street, the address of the bureau's Salt Lake City office. The Utah capital also hosts the Federal Colorado River Basin Forecast Center.

During February, as Salt Lake City prepared for high water, managers at Glen Canyon again released about half the water they could have. In March, the National Weather Service reduced its runoff forecast slightly. Operators cut Glen Canyon's releases by nearly one-third. April found Salt Lake County in the midst of its tenth week of flooding. The ski resort at Vail closed but reopened after eight feet of new snow fell. Copper Mountain extended its ski season by a week. The National Weather Service raised its runoff forecast to 14 percent above normal. But operators at Glen Canyon released no more water than they had in February. By the end of April, Lake Powell stood almost exactly where it had on New Year's Day.

In mid-May, the National Weather Service increased its runoff forecast to 20 percent above normal. For the month, operators again let out half the water they could have. Water ran in the streets of Salt Lake, the late mountain snowmelt streamed its way toward Lake Powell, every rill and rivulet in the West ran high. In spite of months of warning, the bureau had managed Lake Powell so that by the end of May it was 96 percent full and held more water than it had on New Year's Day.

By 1 June, 90,000 cubic feet—675,000 gallons—entered Lake Powell each second, more than twice the water that could leave through the generators and the lowest exit from the dam, the river outlet works. Now the bureau faced its anathema: "spilling" water by having it exit via some other route than through the cash-producing generators. The time had come to turn to Lake Powell's safety valves—the twin spillways, with gates 40 feet wide and 52.5 feet high, located high on the walls of Glen Canyon.

Engineers had designed and tested the spillways but had never had to use them to expel high water. To gauge the hundred-year flood, especially on the Colorado River, it helps to have at least a hundred years of data. When the bureau planned Glen Canyon Dam, it had only about fifty years of accurately gauged flow. It designed Glen Canyon's spillways, generators, and outlet works together to pass 319,000 cubic feet per second (cfs). Would it be enough? The largest flood since the white man had arrived occurred in 1884, but there were no instruments to record it. Using the high water mark on the walls of the Grand Canyon and the readings on gauges upstream and downstream, hydrologists later estimated that the 1884 flood reached 300,000 cfs.[11] Geologic studies suggest that a flood that occurred sometime between 1,200 and 1,600 years ago approached 500,000 cfs.[12]

In 1941, when Lake Mead's spillways had to be used to shunt high water around the dam, cavitation shattered the spillways' concrete linings. From a few feet away, concrete appears smooth, but up close or under a microscope it reveals an irregular, bumpy surface. Water passing at high speed over these bumps creates a momentary vacuum beneath. The vacuum pockets implode like miniature firecrackers and blast out cavities. The shock wave from one cavity leapfrogs downstream and starts another. The resulting runaway erosion can quickly eat through a spillway lining into the bedrock behind. Cavitation can cause concrete and rock debris to break loose and fall, clogging a spillway tunnel. At worst, an entire spillway could crumble, exposing the bedrock beneath to massive erosion. At Glen Canyon, that could have created the fatal connection between the ocean of water in Lake Powell and the cavernous Grand Canyon downstream.

At the time the bureau built Glen Canyon Dam in the late 1950s, the only way to prevent cavitation was to make the spillway linings as bump-free as possible. But a decade later, bureau engineers had found a better solution. In June and July of 1967, high water required the agency to use the tunnel spillway at Yellowtail Dam on Montana's Big Horn River for twenty days. The resulting cavitation gouged a hole in the lining as large as a tractor-trailer rig. Experiments subsequently showed that introducing air into the water traveling through a spillway reduced turbulence and prevented cavitation. The agency built tubes to inject air into the Yellowtail spillways and could have installed similar aerators at Glen Canyon in the late 1960s or early 1970s. It had the technology and the evidence from Yellowtail that the aerators worked. A retrofit into a dry spillway tunnel would cost far less than repairing one after cavitation had damaged it. But the bureau did not take advantage of the opportunity.

W. L. Rusho served as public affairs officer of the Bureau of Reclamation for many years, including the 1983 crisis. In an honest reprise of his career, Rusho asked and answered the obvious question of why designers specified spillway tunnels at Glen Canyon that were almost certain to suffer cavitation damage when used. His response: "A well-managed reservoir should almost *never* spill, and then only for very short periods, after which the cavitation could be repaired."[13] Thus the bureau intended to manage Lake Powell so well that Glen Canyon's spillways would almost never need to be used. Man would control Nature, not the other way round.

By early June 1983, Nature was rapidly gaining the upper hand. Operators gingerly raised the gate on Glen Canyon's left spillway (facing downstream) to release 10,000 cfs. When all appeared well, they doubled the flow. Soon the dam began to shake and shudder like a race car about to throw a wheel. Hundreds of feet downstream, the jet exiting the spillway turned red and entrained large blocks of concrete and sandstone, sure signs of cavitation. Continuing to use the left spillway was risky—cavitation might dislodge debris that would clog the tunnel completely, or water might burrow through the bedrock and undercut the dam. Both could happen. On the other hand, closing the spillway gates risked overflowing the dam.

Bureau officials decided they had to close the left spillway in order to allow crews to inspect the damage. The lake was rising visibly, so they had no time to lose. To compensate for the loss of the left spillway, operators opened the right spillway gate and for the first time began to use the outlet works. A crew of brave inspectors boarded a cart boldly christened, "I Challenge U2." A winch lowered them into the spillway's cavernous maw. Behind and above the inspectors, the steel spillway gate held back the entire volume of Lake Powell. Have faith in U.S. Reclamation indeed!

The inspectors knew where to look for cavitation damage: at the "elbow." When the bureau built Glen Canyon Dam, it piled up a temporary earthen barrier in the floor of the canyon just upstream from the future location of the dam. This cofferdam diverted the river from its bed

Figure 3. Diversion tunnel at Glen Canyon. Once the diversion tunnels were no longer needed, they became the downstream end of the spillways. (Courtesy of Timothy L. Parks, *Images of America: Glen Canyon Dam* [Arcadia, 2004] and Bureau of Reclamation)

and into two huge tunnels carved near the base of each canyon wall. The diversion tunnels discharged the water hundreds of feet downriver, leaving the dam site dry so that construction could proceed. Once contractors finished the dam, the diversion tunnels were no longer needed. As they had done at Hoover Dam, bureau engineers designed the spillway tunnels at Glen Canyon to plunge from high on the canyon wall above the dam down to intersect the old diversion tunnels, which could then serve as the downstream half of the spillways. To understand the geometry, point your upper arm down steeply and extend your forearm horizontally. In between is your elbow; as athletes know, it is a vulnerable joint.

Because the bureau did not have to bore the lower section of the spillways out of solid rock, re-using the diversion tunnels saved a lot of money. But any engineering design for a large project trades off cost against safety. A design that will contain the two-hundred-year flood costs more than one that will survive the hundred-year flood. The decision to use the diversion tunnels saved money but introduced a vulnerability: the elbow. Engineers could have avoided the old tunnels and designed the spillways to run from intakes far enough upstream to exits far enough downstream so that the spillways sloped at a constant low angle—with no elbow. That would have reduced the speed and the turbulence of water flowing through the spillways. On the other hand, drilling the longer tunnels would have cost more and taken longer. If the bureau would manage the reservoir so that the spillways would almost never be used, why bother? Instead, Glen Canyon's spillways descend at 55 degrees, then make a sharp bend where they meet the nearly horizontal older tunnels. To prevent water from flowing back up the old diversion tunnels, toward the lake, several hundred feet of concrete plug the sections above the elbow.

When the operators opened the left spillway tube in June 1983, a thousand tons of water a second crashed down and struck the elbow section. The design was almost certain to produce cavitation, and as Rusho explained, it did. The inspectors found that the imploding water had peeled away several feet of the concrete lining and penetrated the soft, porous Navajo sandstone below. Continuing to use the spillway was obviously dangerous. But the right tube was potentially more so, for it lies closer to the abutment where the dam wedges against the sandstone on the canyon wall. The engineers realized that if they could increase the height of the spillway gates even a few feet, the lake would have room to rise and they could buy time until the peak snowmelt passed. They hit on the ingenious solution of mounting four-by-eight-foot sheets of plywood, bought at a local supply store, to the tops of the spillway gates. Though it seems curious at first thought, the water lapping at the surface of even a giant reservoir has no more sideways force than the handmade waves sloshing the rubber duckies in our bathtubs.

As workers prepared the wooden flashboards, there was no mistaking the pressure the dam was under. Leaks sprang from joints in the outlet works. High pressure popped up manhole covers all over the dam, as if a master magician had levitated them. Everything leaked that could.

The water rumbling through the spillways and the vibrating dam produced a cacophony of sound. Standing in the access tunnels in the dam abutments was like being inside a factory in a rainstorm, as the enormous pressure forced water through the porous sandstone. Those approaching the dam from downstream could hear the noise from four miles away. At two miles, large waves stirred the surface of the river and a violent rainstorm fell from the mist emitted from the spillways. Springs spurted from the sandstone walls of Glen Canyon. Closer to the dam the jets from the spillways began to eat into the protective apron that led back to the base of the dam where the generator releases emerge.[14] To one making the trip upstream, that the largest dam disaster in human history might be under way did not seem far-fetched.

Once the flashboards were in place, operators could open the gates on both spillways. On 19 June, the jet from the left spillway faltered and then stopped. Operators had no choice but to increase the flow, cavitation be damned, hoping to flush out the unknown obstruction. It worked.

Near the end of June, over 100,000 cfs were entering Lake Powell and even higher flows were on the way. Writer T. J. Wolf interviewed the valiant bureau staff responsible for managing Glen Canyon Dam during the crisis and imagined how they might have felt:

> Put yourself in the control room that June 27 morning, when you feel frantic about the left spillway discharge, and you are under orders to reduce the dam's noise and vibration before the turbines start to wobble on their axes and spin themselves into destruction, just before the spillways also self-destruct. You can't shut down the power system, and you dare not shut down the river outlet works. So you turn the dials regulating the left spillway not down, but up—up from the 25,000 cfs that is already performing a tonsillectomy on the left spillway, up to 32,000 cfs. Your other hand revs the right spillway (the one you are really afraid to use) up, up from 10,000 to 15,000 cfs. Counting everything, your dials

tell you the total discharge is 92,000 cfs. What the discharge really is you have no idea. There are no flow gauges down in the bedrock.[15]

A tsunami-like crest of high water rushed past Lee's Ferry a few miles downstream of the dam and fell onto hikers and rafters in the Grand Canyon. The National Park Service had dropped leaflets warning that higher water was on the way and advising those in the canyon to camp high. But at Crystal Rapid, forty-foot, nearly unflippable rubber rafts capsized. The rafters, supported only by their life vests, took the whitewater ride of a lifetime through the next seven rapids. Helicopters evacuated 150 people; many were hurt and one drowned. The surging waters scoured the canyon walls, removing riverine plants and vital sediment.

Back at the dam, the ominous noises continued. Chunks of concrete and sandstone again appeared in the rusty water exiting the left spillway. Operators lowered its discharge and raised the flow through the more dangerous right spillway. Now the plywood sheets became a concern. If the operators closed the spillway gates completely, water flowing over the tops of the sheets (as opposed to merely lapping against them) might wash the sheets away. To prevent that, they decided to replace the wooden sheets with steel ones four feet higher. But the latest forecast predicted that the lake would rise above them.

On the first of July, 122,700 cubic feet of water, the all-time high since Lake Powell began to fill, entered each second. The next day, inflow dropped slightly and the weary staff took hope. But since several days have to pass for the lake at the dam to adjust to high water entering at the head, 186 miles away, there was no time for complacency. James Watt, President Reagan's secretary of the Interior, arrived just in time to proclaim, "The system is working beautifully. Inflows at Glen Canyon are reducing each day."[16] Meanwhile, inside the gallery passageway nearest the right spillway tunnel, drain holes spewed great gulps of air and water.[17] As author Stephen Hannon reports, "Down in the employee dining room, located at the base of the dam adjacent to the left abutment, a worker later said that it sounded like the artillery barrages he had experienced in Viet Nam."[18]And now there was a new worry: operators feared that the motors used to lift the spillway gates might not be powerful enough to open and close them given the extra weight of the steel flashboards.

By 6 July the steel flashboards were in place, allowing operators again to close the left spillway gate and send in an inspection team. The intrepid investigators found broken concrete, tangles of twisted rebar, a hole the size of a large house, and a truck-sized boulder. The amount of debris clogging the floor convinced the inspectors that the erosion had indeed been working toward the dam. In the right spillway they found another large hole and rebar bent like spaghetti.

By mid-July, the lake level climbed to 3,708.34 feet, only seven feet below the dam crest and, had the steel flashboards not been in place, well above the tops of the spillway gates. On 15 July, the lake level dropped one-half inch. It had fallen by that much a few times before, only to rise again. But this time was different. On 16 July, the level dropped another half inch. And another half inch the next day and the day after that. At last, the crisis at Lake Powell was over. The hard work of the dam operators, under the greatest stress imaginable, and the do-it-yourself ingenuity of engineers and contractors, had averted a calamity. But there was no time to lose: the spillways were shattered and in only nine months, as the next spring's snowmelt arrived, Lake Powell would again start to rise.

As soon as the lake level fell below the spillway gates in the late summer of 1983, crews entered the spillways and began to fill the holes in the linings, a job that took 2,300 cubic yards of concrete. They cut air slots in the spillway tunnels to prevent cavitation, as they had done at Yellowtail sixteen years before. The next spring, just before the snowmelt began to arrive, operators brought the lake to within nine feet of its April 1983 level. Inflows during the second half of May and early June 1984 topped 100,000 cfs, again requiring use of the spillways, but there was no sign of cavitation.

Afterward, Commissioner Broadbent and other bureau officials said that nothing could have been done to prevent the crisis and in any case, there had been no serious danger. But other bureau staff disagreed. "How close did we come to losing Glen Canyon in 1983?" asked a future reclamation commissioner. "We came a hell of a lot closer than many people know. I mean, it was digging sandstone when they finally got a handle on it."[19]

The bureau presented its official history of the emergency in a film. According to the narration, the crisis began "in May [when] heavy snowstorms hit the high country. It was cold then, but the heat of summer cut like a hot knife and the heavy snowmelt was on." The subsequent repair of the spillways and the belated cutting of the air slots was a "monumental accomplishment"; more than that, it was a "victory for the human spirit [and] for the leaders who cut through the red tape."[20]

Broadbent blamed "a faulty computer model that needs some revision."[21] As James Udall pointed out, this was the perfect bureaucratic cop-out. It succinctly "supplies the scapegoat (the computer), the cause (its faults) and the cure (its revision)."[22] And no bureau employee, indeed, no human being, could be found at fault.

No geologist is surprised that the Colorado River threatened Glen Canyon Dam—the river has removed every grain of rock that once occupied each of its many canyons, including Glen and Grand. The Grand Canyon is roughly three hundred miles long, fifteen miles wide, and one mile deep. This means that the river and its tributaries have excavated an average of 125 million tons of solid rock from the Grand Canyon—each year for the last five million years or so. Not only that, the Colorado has removed even larger dams than Glen Canyon and Hoover. In the last half a million years, the river has blasted to smithereens a whole series of hard lava dams in the Toroweap section of the Grand Canyon, strewing their remnants downstream for eighty miles.

Held in our hand, water is a puny thing. One could spend a lifetime dribbling it over a block of basalt or concrete with no noticeable effect. Yet give it a slope of several feet per mile, increase its volume to thousands of cubic feet per second, and wait for only an eyeblink of the deep time of geology. Then rock and concrete turn out to be puny.

In 1983, not on a geologic time scale, but on ours, the Colorado came close to removing the latest obstacle in its path. Less than two decades later, as the twenty-first century began, Glen Canyon Dam and Lake Powell face a new threat. Rather than too much water, this time the problem is too little. Unlike earlier floods and droughts, this one gives no signs of ending anytime soon.

TWO

Playing Dice with Nature

BY 1999, A PREDOMINANCE of wet years had brought Lake Powell to full pool. Then, just as the Colorado River system of dams and reservoirs appeared poised to begin a successful second century of reclamation, Lake Powell began to fall and fall fast. The flow of the Colorado River dropped so rapidly that in 2002, the reservoir received only one-quarter of its average inflow, the second lowest on record. By early 2005, the reservoir was only one-third full. Had the drought continued at the same pace for two more years, power production at Glen Canyon would have ended. Another year and the lake would have fallen to the elevation of the dam's outlet works, the lowest exit—at the level called dead pool. Even then, Lake Powell would have a surface area of over 20,000 acres, causing it to evaporate more than 100,000 acre-feet of water each year. At dead pool, only two of Lake Powell's five boat ramps would reach the lake surface. Once on the lake, boaters would find their favorite moorings and campsites scores of feet above the water line. In order to meet downstream delivery contracts, operators would have to release through the outlet works every drop of water that entered Lake Powell. To speak of managing the reservoir would no longer make sense; Nature would have reassumed management.

The twenty-first-century drought struck so suddenly that it fooled even the experts. In 1995, the *Water Resources Bulletin* published a study of the predicted effects of long-term drought on the Southwest.[1] The report brought together the leading authorities on the hydrology, meteorology, climate, and dendrochronology of the Colorado River Basin. Trying to identify the worst case, the experts estimated the frequency of long-term drought and how it would affect the flow of the river, the amount of water in reservoirs, water markets, and so on. Reality caught the scientists as much by surprise as the 1983 flood had caught the Bureau of Reclamation. The flow of the Colorado River during the twenty-first century dropped so much faster than the experts thought possible that by 2004, Lakes Powell and Mead together held 20 MAF less than their worst-case forecast. It was as though in four years a reservoir nearly as large as Lake Powell had simply vanished. That Lake Powell could ever reach dead pool—only three years away in 2005—had been deemed "unrealistic" by scientists. They assumed that a truly severe drought would occur years in the future, preceded by several less-severe ones that would allow water managers time to adjust. Instead, four years after the report appeared, the worst drought in the last century and one of the worst in the last five centuries began. One reason the predictions may have fallen so far short of reality is that in 1995 the scientists did not have enough information to factor global warming into their analysis.

The authors of the 1995 study thought that a harsh drought would increase the price of water and lead to less consumption. Instead, business as usual prevailed. As the level of Lake Powell fell, power managers continued to release the required amount of water, though it generated less power as the hydraulic head on the generators fell. The managers really had no choice. The entangled web of statutes, contracts, agreements, treaties, administrative actions, court decisions, and regulations that collectively make up the "Law of the River" prevents a rapid response to drought and, as we will likely find out the hard way, any effective response.

The falling lake exposed an ugly, white bathtub ring where high water had deposited minerals onto the canyon walls. Soon the pale band rose ten stories high, compromising Lake Powell's beautiful vistas of sapphire water set against ocher sandstone bluffs. The low water stranded marinas and boat ramps high above the lake. The National Park Service, chasing the receding lake, had to extend the boat ramp at Bullfrog Marina so far out that had the ramp been level, a small plane could have landed on it.

When Congress approved Glen Canyon Dam in the 1950s, the use of its reservoir for recreation received little attention. Yet Lake Powell and the surrounding Glen Canyon National Recreation Area soon became one of the major vacation spots in the United States, in some years drawing more visitors than Yellowstone National Park. At the peak in 1993, almost 3.6 million people visited Glen Canyon. Then, well before the drought, the number of visitors began to decline and by 2004 had fallen by half. Gasoline prices and post-2001 fears of terrorist attacks likely had something to do with the decline, but so did the unsightly bathtub ring and the inability of boaters to launch and, once on the water, to reach their favorite shorelines and campsites.

The lower Lake Powell fell, the more one group took heart. Reclamation commissioner Floyd Dominy had named the lake for the first person to run the length of the Colorado River canyons, Major John Wesley Powell, who had been dead for sixty years and could not object. Dominy's bureau christened the dam itself for something that was not yet dead but soon would be, the canyon that Lake Powell would submerge. Powell and others had judged Glen Canyon to be more beautiful than even the Grand Canyon. Author Wallace Stegner agreed and spoke for many: "Awe was never Glen Canyon's province," he wrote. "That is for the Grand Canyon. Glen Canyon was for delight."[2]

As Lake Powell fell, Glen Canyon began to re-emerge. The rafters who had seen it four decades before, who had camped in its grottoes, floated into its alcoves, and sought solace in its ethereal natural cathedrals, had thought that they would never see Glen Canyon again. But as the lake shrank in the summer and fall of 2004, hopes rose that people might even be able to enter Glen Canyon's icon: the Cathedral in the Desert. By late March 2005, the lake level had fallen just enough to allow a few determined lovers of Glen Canyon to reach the Cathedral by boat and photograph its vaulting walls and arched ceiling. But they had to be quick—the resurrection would not last long.

During the winter of 2004–05, from Malibu to Denver, heavy rain fell along the coast and in the valleys, while snow piled high in the Rockies. The deluge set records in California and far inland. By mid-March, the snowpack in the Rocky Mountain drainages that feed the Colorado River stood at 110 percent of average. Since melting snow provides more than 75 percent of the water in the river, Lake Powell looked set to regain much of its lost elevation. Sure enough, the reservoir rose steadily, again submerging Glen Canyon and drowning the hopes of its devotees. By mid-July, the lake crested at 3,608 feet, some fifty feet above its April 2005 low but still nearly ninety feet below its 1999 high.

Depending on whether one thought the drought was over, Lake Powell was either half full or half empty. It was natural to expect that 2006 would answer which. The year began well enough, with January and February snowpack levels matching or topping historic averages. Then temperatures rose and the snowpack shrank until by mid-March it stood at 90 percent of average. But the bad news was only temporary. In late March heavy snow fell and by the end of the first week in April, snowpack had risen to 107 percent of average, nearly as high as the year before. The Bureau of Reclamation forecast that runoff into Lake Powell would be close to normal. The Natural Resources Conservation Agency projected that the elevation of Lake Powell would rise to 3,636 feet, about 70 percent full. Then to the forecaster's surprise, the snowpack began to vanish. By mid-May, it had shriveled to less than half its historic average.

Inflow to Lake Powell remained close to average until early June, then in response to the reduced snowmelt, it plunged and stayed low throughout the summer. In June and July of an average year, the Colorado River carries 3.5 million acre-feet of water to Lake Powell. But during the same two months of 2006, only 2.2 MAF entered, 1.3 MAF less. How much water is that? One-tenth of Lake Powell's current volume, twice the annual consumption of the city of Los Angeles, one-third the total consumption of the four upper basin states, half of Arizona's allocation of Colorado River water, and so on. How long did it take the Colorado River system to lose that much water? Sixty-one days.

Lake Powell eventually peaked at 3,611 feet, well below the forecasts. On the east side of the Rockies, a similar loss took place. Between April and May, many of the snowpack sites that feed the Platte River lost over 50 percent of their volume—and some lost 100 percent.[3]

Where did the potential Rocky Mountain snowmelt go if it did not enter the streams? A clue is that the first six months of 2006 were among the warmest on record in the United States. Higher temperatures obviously cause more and faster melting, but does not the same amount of meltwater flow downstream, only sooner? Evidently not. Warmer temperatures not only melt snow, they cause it to sublimate—to pass directly from solid into vapor without going through the liquid phase. Think of clouds of carbon dioxide vapor streaming from a block of dry ice. Water vapor from sublimating snow wafts away on the wind, to condense and descend as rain somewhere, just not necessarily in the river basin where it originated. Higher temperatures also dry out soils, causing them to absorb more of the snowmelt. Some combination of these effects led to the Houdini-like great snowmelt escape of 2006.

———

These first two chapters show how in only seventeen years, two major and opposite crises on the Colorado River threatened the Southwest. First came the flood, then came the drought. Both caught even the experts by surprise, revealing that in deciding to store twenty trillion gallons of Colorado River water behind concrete arch dams, then allowing a civilization to become dependent on that water, we did not know what we were doing. We gambled and got away with it—for now. But Nature is an implacable opponent, with unlimited time and energy. Keep rolling the dice with her, and eventually you will lose.

As the twenty-first-century drought hung on, westerners naturally wanted to know when it would end. Always in the past, waiting had proved a workable management strategy. Floodwaters eventually receded; wet years always arrived to break droughts and fill depleted reservoirs. By adding flashboards to the top of Glen Canyon's spillways in 1983, the bureau bought time until the wave of high water passed. Through 2000, 2001, 2002, 2003, and 2004, water officials waited. Finally, in 2005, three years away from dead pool on Lake Powell, the rain and snowmelt arrived. But as it registers that global warming has already begun and is going to get worse for the rest of this century, some have begun to realize that this and future droughts may not be "over," in the twentieth-century sense. Including the relatively wet 2005, the average inflow to Lake Powell during the first eight years of the twenty-first century is down by an average of 40 percent. To restore the twentieth-century average by, say, 2025, would require that flows for the next seventeen years average 20 percent *above* normal, a sequence we can safely bet against. One of the conclusions of this book is that during the rest of this century, there will often be only enough water in the Colorado River system to maintain one large reservoir, not two. Even one reservoir will sometimes be no more than half full. As one climate scientist put it in April 2007, "The Dust Bowl and 1950s droughts will, within the coming years to decades, become the new climatology of the American southwest."[4] Not just the weather, but the very climate itself, has begun to change.

Though the scientific consensus that humans are the principal cause of global warming grows stronger every month, to forecast trouble ahead for the Colorado River basin, one does not have to get into causes. Global warming is well under way in the western United States and, regardless of the cause, it will continue for decades. Nothing that the West can do in the short run to reduce greenhouse gas emissions will have any effect on the Colorado River. The supply of its water is going to fall; our best bet is to focus on demand. But that will take a degree of political will and leadership seldom seen in hydraulic societies.

The "Concrete Pyramid"—the Bureau of Reclamation and its companion agency, the Army Corps of Engineers; pork-barreling western politicians; mega-construction companies; and agribusinesses on welfare—already proposes to combat global warming and solve the water crisis of the twenty-first century by building more dams and reservoirs.[5] Before we accede, Americans need to understand the answers to several questions. Why did our government dam nearly every river in the West, some a dozen times or more? Why were dams built even though the associated irrigation projects were obvious money-losers? Why, within a decade or two of the launching of the United States Reclamation Service in 1902, were every one of its founding principles betrayed?

Most important in the long run, what do the dam-builders have in mind when reservoirs fill with silt, as they inevitably must unless the law of gravity is repealed. What do we do when across the West are spread not beautiful blue-water lakes, but a hundred million acre-feet of mud, some of it laced with toxins? Where then will our successors get their water?

Millennia of human history demonstrate that reclamation has insidious internal flaws. Yet in the go-go years of twentieth-century dam-building, America acted as though those flaws would not apply. But nothing is new under the sun: like Las Vegas revelers pulling on one-armed bandits while fountains spray water into the dry desert air, hydraulic societies have always deluded themselves into believing they could gamble with Nature and win.

PART TWO

River of Empire

THREE

Appointment in Samarra

IN A.D. 1255, THE GREAT KHAN, Mongke, dispatched his brother Hulagu—
they were two of the three grandsons of Genghis—to subjugate the
remaining Muslim states in southern Asia.[1] One foe, the fearsome Assassin
sect, surrendered without a fight. At the head of the largest Mongol army
in history, in 1258 Hulagu arrived at the Nahrwan region east of Baghdad.
He had come to sack the city, whose caliph had refused to provide Hulagu
with fresh troops. The area surrounding the Nahrwan River had been a
rich agricultural region for thousands of years. But when the Mongol gen-
eralissimo gazed out upon it, all that remained were a few dying towns
spread among a vast network of silt mounds. The only water came from a
thin trickle in the main canal. Historians once blamed the Mongols for the
destruction of Nahrwan, but of that particular charge they were innocent.
By the time Hulagu and his horde arrived, irrigation's inevitable ills had
already done the job.

The region Hulagu bestrode had been irrigated for thousands of years.
One after another, the cities of the Fertile Crescent—Ur, Ashur, Nimrud,
Nineveh, Samarra, and Babylon—had risen along the riverbanks of
Mesopotamia, and fallen. As far back as the sixth millennium B.C., the earliest known civilization, the Sumerians, developed a sophisticated network
of irrigation canals and controls. Centered on city-states like Ur, which at
its height in the third millennium had 250,000 people, the Sumerians

developed a rich culture. They invented writing, the wheel, the sixty-second minute, the sixty-minute hour, the yoke, the first code of law, and the city-state itself. Surely the Sumerians believed that their civilization—the highest the world had achieved—would last forever. Yet when Babylon rose in the eighteenth century B.C., it built atop the ruins of a vanished Sumer.

Irrigation societies have three enemies: silt, salt, and their own internal and political weaknesses. First, silt. Irrigators must slow a stream enough to divert its water safely through their irrigation works and into canals. But as water slows, it loses energy and its ability to carry sediment. Silt drops out, clogging canals and raising both the bed of the river and the level of the fields. In one five-hundred-year period, the fields of Nahrwan rose by three feet.[2] On the other hand, if canals slope too steeply, the rushing water may erode and destroy them and the associated irrigation works. The early irrigators had to master a delicate balance between too much slope and too little.

If silt does not defeat an irrigation society, salt eventually will. All natural waters, even rainwater, contain dissolved salts. When irrigators flood their fields, plant roots take up some of the water, but three-fourths evaporates from soil and plant leaves. The salt stays behind and steadily builds up in the soil. Some salt is essential, in part for the trace elements that accompany it, but too much is poison. The saltier the soil and groundwater, the more energy plants must expend extracting the moisture they need and the less energy they have left for growth. In 2400 B.C., the fields of Sumer yielded about thirty bushels of grain per acre, surprisingly close to modern wheat production. Four hundred years later, yield had slipped to seventeen bushels per acre, and four hundred years after that, to ten.[3]

Declining grain yields in ancient Mesopotamia gave rise to a vicious feedback. Farmers planted and irrigated more fields, requiring more canals, which led to increased siltation and higher salinity over larger areas. More and more workers and organizational apparatus had to be devoted to clearing and rebuilding the choked irrigation works. Finally the task became so large and complex that only the most skillful and diligent managers could achieve it, and then only for so long. The administrators of the Nahrwan region succeeded for millennia in bringing water to their people, but finally they failed and their society vanished. Modern archeologists have no trouble identifying the ancient silt mounds left over from thousands of years of canal cleaning in Mesopotamia.[4]

As if silt and salt were not enough, add Middle Eastern politics and adventure. A stable irrigation system requires a hierarchy, strong central control, and eternal vigilance. Sumerian authorities became more concerned with wars and intrigues than with the maintenance of their irrigation system; all went down together. More than a thousand years after Sumer, wars, burdensome taxation, and inattention by its Persian conquerors allowed the canals of Babylon to deteriorate, opening the way for Alexander. Yet even the great commander could not avoid the inevitable consequences of irrigation. In the fourth century B.C., Alexander had to put 10,000 people to work for three months cleaning out a single, silt-clogged canal.[5]

The Tigris and Euphrates flow roughly parallel for hundreds of miles in a northwest-southeast direction. In order to transfer water, ancient irrigators dug canals to connect the two rivers. The basins in between filled with water, which then evaporated, raising salt levels and poisoning the fields. By the seventh century A.D., the only way to get rid of the salt was to strip off the saline upper layer by hand. Only slaves could be made to suffer such inhuman working conditions. "By the sixteenth century," writes Postel, "the Fertile Crescent of Mesopotamia, from which human civilization had sprung and reached unprecedented heights, was little more than a salty wasteland."[6]

Donald Worster sums up the lesson of history:

Reclamation . . . cannot be indefinitely sustained. As the irrigation system approaches its maximum efficiency, as rivers get moved around with more and more thorough, consummate skill, the system begins to grow increasingly vulnerable, subject to a thousand ills that eventually bring about its decline. Despite all efforts to save the system, it breaks down here, then there, then everywhere.[7]

Eventually, silt, salt, and politics catch up with each irrigation society. Modern Egypt escaped the triple threat for millennia, but it may prove to have no permanent immunity. The ancient Egyptians had begun to irrigate even earlier than the peoples of Mesopotamia. Around 8000 B.C., climate change, perhaps coupled with overgrazing, converted what had been tribal pastoral lands into the modern Sahara desert. That forced nomadic tribes to the banks of the Nile, where they settled and where the cities of ancient Egypt rose. Like the enormous salmon runs that once miraculously reappeared each year in the Pacific Northwest to sustain Indian populations, the Nile offered up its own annual gift from the gods. In the middle of the

boiling hot and relentless Egyptian summer, the muddy Nile would myste-
riously rise and spill over its banks onto the adjacent floodplain. When the
floodwaters receded, they left behind a thin layer of rich topsoil. "Egypt is
the gift of the Nile," wrote Herodotus in the fifth century B.C.

But in some years the river failed to flood; in others the waters ran in
flood and washed away the fertile silt, as well as excess salt. Being able to
predict the behavior of the Nile became critical to early Egyptian agriculture. Those who could do so inevitably gained power, leading the priesthood to take over the task. By measuring the height of the Nile at Aswan
with a crude scale called the Nilometer—said to be the first scientific
instrument—the priests were able to predict how high the river would rise
downstream.

Since the ancient Egyptians did not have to control and direct water, silt
was a boon rather than a burden. They cleverly allowed floodwaters to
flow into basins and remain until it saturated the soil, after which they
drained the basins. That kept the water table several meters below the surface, preventing salt from moving up into the root zone.[8]

Thousands of years after salt, silt, wars, and political intrigue had
destroyed Sumer and Babylon, the natural irrigation system of the Nile
sustained Egypt and allowed it to avoid the fate of other ancient irrigation
societies. Egypt survived not only its own political machinations but also
conquest by Persians, Greeks, Romans, Turks, and Arabs.[9] Then came the
rise of nationalism, the Cold War, and the new religion of mega-dams and
its priesthood. In two generations, a natural system that sustained Egypt
for eight millennia has been brought to the verge of failure.

In July 1952, a group of Army officers led by Colonel Gamal Abdel
Nasser overthrew corrupt King Farouk. To dramatize their break with the
past, the Revolutionary Command Council sought a grand new project—
ideally something on the scale of the pyramids. Their solution was a giant
dam at Aswan that would capture twice the yearly volume of the Nile and
replace the annual flood cycle with steady and dependable releases, allowing two or perhaps three crop rotations a year. The dam would even generate enough hydropower to pay for its construction. Those were the
benefits; with the collusion of the World Bank, the costs were shamefully
underestimated or ignored. When the Suez crisis led the United States to
withdraw from the project, the Soviet Union stepped in with money and
construction expertise. The Aswan High Dam flooded much of lower
Nubia, forcing over 90,000 people from their homes and drowning
unique archeological ruins.

Fifty years later, the true costs of the dam at Aswan are becoming evident. Before the dam, the Nile deposited about 10 percent of its sediment on Egyptian fields, sending the rest on to the Nile Delta. After the dam, nearly 100 percent of the sediment began to pile up behind it, starving downstream fields and the Nile Delta. One author estimates that Aswan Reservoir has about 150 years before silt fills it, but that does not take into account potential reductions in the volume of the lake due to global warming.[10] Vast quantities of chemicals have to be added to replace those that Nile silt provided free, making Egypt one of the world's heaviest users of fertilizer, with its accompanying pollution. The fertilizer plants obtain their electrical power from Aswan Dam.

Evaporation of irrigation water annually deposits about half a ton of salt on each Egyptian acre, reducing agricultural yields. Egypt has spent $2 billion to drain salty water from five million acres, much of the money provided by the World Bank. Following the tradition of past irrigation societies, as salt and silt compromised production, the Egyptian authorities annexed and irrigated more land. But the new fields never met their productivity expectations.

Historically the Nile Delta has been among the most fertile regions in the world. The downstream edge of its fan-shaped form curves along 240 kilometers of the Egyptian coast from Alexandria to Port Said. A delta is the ideal natural equilibrium system, balancing the supply of sediment from an incoming river with the demand from sea waves. With the Aswan Dam cutting off the supply of sediment and demand from the sea constant, the Nile Delta has been in steady retreat, each year losing more of the world's richest soil. In the time of Pliny the Elder, the Nile Delta had seven distributaries; now it has two. A village once situated at the mouth of one of the main channels is now two kilometers out to sea.[11] Like other large rivers today—the Colorado, the Rio Grande, the Murray-Darling, and the Yellow—the Nile often runs out of water well short of the sea.

Hydrologists estimate that if current trends continue, by 2020 the annual demand for water in Egypt will exceed the natural supply of the Nile by 11 MAF. The estimate does not include the effects of global warming, which could decrease runoff into Lake Nasser considerably more. According to one calculation, a decline in rainfall of 10 percent would reduce runoff in the watershed of the Blue Nile by 35 percent.[12] The Nile crosses ten African countries, more than any river on the continent. Egypt and Sudan have rights to all the water in the Nile, portending trouble for some of the desperately poor countries upstream.

Already, a war is being fought in Sudan with water as one of its root causes.

Irrigation began in North America thousands of years after the Sumerians. The Spanish explorers who arrived in the Tucson basin of Arizona in the 1600s met small bands of Pima and Tohono O'odham Indians. The two tribes were themselves recent arrivals, having reoccupied lands abandoned by an earlier society. The Indians called these ancients the *huhu-kam,* a term that means "all used up."[13] The Hohokam, as we know them, were the most successful irrigators and agriculturalists in prehistoric North America. But like the Anasazi, their contemporaries to the north, just when Hohokam society was at its height, its people vanished.

When Americans came to central Arizona and began to dig canals in the mid-nineteenth century, they usually found that they were excavating on top of an ancient Hohokam canal. Topography and upstream water supply dictate the best places for canals, and none of those factors had changed since Hohokam times.

As writer Craig Childs points out, until recently Phoenix built laterally rather than vertically.[14] Few homes had basements and there were no skyscrapers. Today, tall buildings, which require substantial excavation, rise over central Phoenix and the city plans many more. The excavations for the new skyscrapers have uncovered previously unknown Hohokam ruins; the Native American Graves and Repatriation Act requires that they be protected. As a result, more has been learned about the Hohokam in the last several years than had been discovered up until then.

Digging for the new convention center in Phoenix uncovered artifacts that show that the first Hohokam people arrived in central Arizona as far back as three thousand years ago. Once they learned how to control and manipulate water, nothing prevented them from expanding to the maximum carrying capacity of the land and water. Eventually their settlements stretched from north of Flagstaff 250 miles south to Tucson.

The Hohokam laid stones and brush across streams to make check dams, jerry-rigged contrivances that trap silt and hold back enough water to provide a modest, semi-permanent supply. The next big flood would wash the fragile dams away, but they were easily rebuilt. Using sharpened sticks and primitive hoes—a pointed rock with a handle tied on—the Hohokam dug over nine hundred miles of irrigation canals and connectors that directed the spring runoff of the Gila and Salt rivers to their fields. To strike a balance between erosion and siltation, Hohokam canals ran with carefully calibrated drops of a few feet per mile. One of their largest, at Pueblo Grande, watered 10,000 acres.[15]

Hohokam ingenuity went beyond manipulating water. Using rafters and I-beams, they constructed four-story houses. They threw pots, worked with shells obtained by trading with Mexico, and played ball games in enclosed courts. Their waste piles show little evidence that they ate meat, suggesting that irrigated agriculture satisfied their needs. At the height of Hohokam society, 40,000 lived where today Arizonans survive with air-conditioning and by consuming over two hundred gallons of water per person per day, water pumped uphill from the Colorado River, hundreds of miles away.

Hohokam society flourished for centuries, then suddenly, in the mid-fifteenth century, it collapsed. Scientists have not been sure why, but new evidence is beginning to provide answers. A tribe living in a desert, dependent on a complex but still primitive irrigation system, was vulnerable both to high water and to low. As long as floods and drought did not last too long or occur too frequently, though individuals and families might perish, the society itself could survive. One researcher has used tree rings and other evidence to discover that around A.D. 1190, weather patterns in central Arizona became more changeable. Floods and droughts along the Salt River happened more often, denying farmers time to recover before the next extreme arrived. A recent tree-ring study found that between 1130 and 1154, the Colorado River had the lowest flow of any twenty-five-year period since 762.[16] The Hohokam death rate began to rise at about the same time, to sixty-six out of every thousand people annually, compared with roughly one-tenth that rate in the United States today.

The Hohokam collapse demonstrates that a desert society that expands to the limit of the carrying capacity of land and water balances on the thin edge of survival. When the inevitable change in climate arrives to perturb the balance, the society has no margin for safety. Usually it responds by impounding more water, planting more crops, and building more canals. But if the problem is too little water for too long, none of these remedies help. A society must respond to a prolonged drought by reducing consumption until demand matches supply, or by migrating.

The Hohokam were skilled enough at controlling and distributing water to live comfortably in the desert for a thousand years—then they disappeared. As Babylon built on the ruins of Sumer, so modern-day thirsty

Phoenix—the bird that rose from its own ashes—sits upon the canal network of the vanished Hohokam.

———

Four hundred years after the Hohokam disappeared and several hundred miles to the north, a small group of Americans tried to make the desert bloom. Their story begins in 1827 in a town in western New York, where a young man named Joseph Smith, Jr., reported that an angel named Moroni had presented him with a set of engraved golden plates. After God showed him how to translate the engravings, Smith published the result as the Book of Mormon and founded a church.

Smith and his followers moved to Illinois and built a new town. There they became so influential that Smith announced he would run for president of the United States. When dissenters among his flock attacked him in their newspaper, Smith had their printing press destroyed. For this and other alleged crimes, and in part for his own protection, the state put Smith in jail. But the protection proved insufficient, for a mob broke in and murdered the Mormon leader. When the violence continued, Brigham Young, one of Smith's elected successors, decided to lead his people far to the West. There they would find an isolated spot to establish their own government and practice their religion without fear of persecution.

After trekking for well over a thousand miles, in July 1847 the Mormon vanguard reached the valley of the Great Salt Lake. As he gazed for the first time on the dried bed of an old sea that would be the Mormon's new homeland, Young proclaimed, "This is the place." They christened their new state Deseret, a name taken from the Book of Mormon that means honeybee. The indefatigable insect became the Mormon symbol.

The Mormons were easterners with no tradition of irrigation. But they had studied the subject and visited sites in the Southwest where Indians and Mexicans irrigated. Emissaries had even journeyed to the Middle East to examine irrigation practices there. Most importantly, Young's flock had organized itself to begin irrigating the moment it reached the promised land.

The advance party arrived early in the morning and set to work. By 11:30 A.M. they had staked out a plot and by noon had plowed the first furrow. By 2:00 P.M. the dam and ditches to lead water to the fields were ready. By the end of the next day, the brethren had planted a five-acre patch of potatoes and turned water onto it. In less than thirty-six hours, while another group might have been arguing whether a better site did not lie just over the hill, Mormons were irrigating.

Mormon tools and techniques at first were little advanced beyond those of the Hohokam. Planks bolted together in the shape of the letter A made a plow; a pan of water became a level. The Mormons made their first dam from dirt and rocks scraped and piled together—a Hohokam check dam.

Unlike the Hohokam, who for centuries had no reason to move, the Mormons felt a divine mission to spread the gospel of desert conquest. Young sent Mormon scouts and parties to build new settlements in the most remote and inhospitable regions of the barren, arid Colorado Plateau. Some of their routes, hacked out of solid rock and seemingly impossible to traverse even today, defy our belief—but did not defy theirs. Once they arrived at their remote destination, just as they had done in the Salt Lake Valley, Mormon pioneers wasted no time digging the first irrigation ditches. By 1910, Utahans irrigated nearly one million acres.[17]

Deseret had many advantages for the Mormons. The Salt Lake Valley was isolated and in the beginning at least, no other whites aspired to live there. The nearby snowcapped peaks of the Wasatch Range promised a perpetual supply of water. As important, the Mormons organized themselves into a tight-knit social organization with a common purpose and a clear hierarchy. To travel across nearly half the continent under the most severe conditions, not knowing where you would settle or that you would be able to survive when you did, watching members of your family die along the way, required faith, stamina, commitment, strong leadership, and equally strong followership. Once in the Salt Lake Valley, their organization and hierarchy allowed the Mormons to make crucial decisions with the authority that history shows is necessary if an irrigation society is to succeed. The church decided where to capture water and how to distribute it. There was no private ownership: water belonged to the people in common. In mid-nineteenth-century America, these were radical, even dangerous notions, a kind of theological socialism that would have been anathema anywhere else in the country. Like any irrigators, the Mormons had to deal with silt and salt. Yet it did appear that they had overcome the political ills that crippled earlier hydraulic societies. As we will see, Mormon irrigation became subsumed into the reclamation movement of the twentieth century, leaving us unable to judge how long the experiment would have lasted on its own.

One western explorer spent a good deal of time with the Mormons. Their model, in which small groups of farmers held land in common and built modest, sustainable irrigation works, impressed him mightily. Though his father had been an evangelical preacher, John Wesley Powell was not openly devout, ordering his Colorado River crew to sail on, even on Sunday. But he respected the Mormons and the feeling was mutual. Powell came to believe that the Mormon example could be scaled up to serve the nation and rescue the hordes of settlers about to be lured west by the false promises of politicians and boosters who assured them that rain would follow the plow. Powell soon came to have more influence on science and water policy than any person in the nation, though it would prove insufficient to allow him to achieve his dream of a sustainable irrigation society in the American West.

FOUR

One Simple Fact

AS LATE AS THE 1860s, geography divided America into three parts. A populous eastern region grew less so to the west until, somewhere out near the 100th meridian, lay the frontier. On the Pacific coast, a few fledgling cities had sprung up and just inland, gold mining and agricultural outposts had a foothold; beyond them to the east lay the nearly impassable Sierras. In between the settled eastern and western regions lay two-fifths of the country, remaining much as Lewis and Clark had found it five decades before, the province of Indians and mountain men.

Most eastern politicians of the day had never seen the middle and western sections of the country. How could they have grasped the distances? Lewis and Clark traveled 3,700 miles—one way—four times the distance from Washington, D.C. to St. Louis. It stood to reason that in the vast expanse bordered by the two settled regions, weather and climate would vary to extremes. Farming methods that succeeded in the East could not be expected to work as well, if at all, in a section so dry that from the time of explorers Zebulon Pike and Stephen Long in the early 1800s, it had been known as the Great American Desert.

America could have let Nature have her way and left much of the region west of the 100th meridian unpopulated save for a few scattered oases. But for the leaders of the nation to admit that a large fraction of the country might not be arable was to give up on several dreams at once: westward expansion, the Jeffersonian ideal of the sturdy yeoman farmer carving a living out of wilderness, the biblical mandate to make the desert blossom, and a union of states. Had the West remained as the explorers and Indians found it, in effect the region would have seceded from America's vision of progress. That was no more acceptable than to allow part of the Union to secede for some other reason. As John Widtsoe, Mormon leader and early twentieth-century irrigation expert would later put it, without the foresight of its statesmen the nation would have been "two strips of country, with a desert between."[1] Aridity was the common enemy, and the weapon with which to defeat it and unite the country was water. Irrigate the arid lands and they would become as fecund and prosperous as any, allowing America to fulfill its God-granted destiny. To say otherwise took a man who was unusually brave, if not downright foolhardy. Certainly he had no future as a politician or a preacher.

John Wesley Powell was neither, but he was a prophet. During his day he was best known for his thrilling adventure running the canyons of the Colorado River in wooden boats. From there he went on to help found and lead the United States Geological Survey and the Bureau of Ethnology. At Shiloh in April 1862, a Confederate minié ball cost him his right arm at the elbow, an injury that would have guaranteed an honorable discharge in any war. But after a brief recuperation, Powell returned to battle and served until the Union victory. His final rank was Major, a title he proudly wore for the rest of his life. His superb biographers, Wallace Stegner and Donald Worster, show how, even though he never ran for office, Powell became one of the most influential men in Washington, especially where the West was concerned. He had been a farmer, a scientist, a western explorer, and a friend of Mormons and Indians. Collectively these experiences made him the best-prepared person in the country to understand the West's potential for irrigated agriculture, as well as its limitations. Powell should have been lauded and remembered for having shown America the way to sustainable, small-scale irrigation in the West. But until Stegner's biography of 1954, Powell and his philosophy had been largely forgotten, submerged like his canyons in the era of reclamation.

The United States acquired vast tracts of western lands long before anyone but explorers and mountain men had seen them. In 1803 President Thomas Jefferson brought into the federal domain the largest single

Figure 4. John Wesley Powell, age 39, with Taugu, Paiute chief. Few western explorers would have posed with an Indian in this collegial manner. (U.S. Geological Survey)

block, the 1.5 billion acre Louisiana Purchase. For much of the nineteenth century, Congress tried to entice settlers onto the western territories. Some politicians thought pioneers should have to buy land; others argued that they should get it for free. Eastern landowners opposed a giveaway, fearing their workers would abandon them. Southerners worried that settlers in free territories would oppose slavery and bring new abolitionist states into the Union.

In 1861, the South seceded and its opinion and votes no longer counted. Within a year, Congress passed the Homestead Act. In return for a small filing fee and a promise to occupy the land for five years and make modest improvements, any head-of-family could receive title to 160 acres of the public domain. The act did not provide loans nor did it address how the homesteader would get water.

Some public lands were barren desert, others were impenetrable swamp. To entice settlers onto such inhospitable terrain, the Timber Culture Act of 1873 offered 160 acres free provided that forty of them were planted with trees— no residence required. The obligation to plant trees reflected the belief—at least, no one had disproved it—that trees cause rain. Under the Desert Land Act of 1877, anyone who would irrigate within three years could buy 640 acres at $1.25 per acre, proof of irrigation left to the buyer. The Swampland Act of 1849 offered incentives for those who would drain wetlands.

Collectively, this legislation presumed that the federal government could manage a huge system of land giveaways, sales, set-asides for military posts, evidence of irrigation or of navigation, and so on. All the while, various protections were supposed to prevent speculators and cheats from taking advantage. It was asking too much. The General Land Office, responsible for the whole house of cards, failed so miserably that Donald Worster called it "one of the poorest bureaucracies ever created."[2] Stegner wrote that the Homestead Act and its supplements "could not have promoted land monopoly and corruption more efficiently if they had been expressly designed for that purpose."[3]

By the mid-1870s, the stench of land corruption was strong, though it was but one of many offensive odors wafting from the crooked White House of Ulysses S. Grant. Major Powell was the exception—as honest a man as could be found in public life in this period. He appears to have done nothing, other than write some overdramatized and not strictly accurate accounts of his river adventures, for personal gain. To a man of Powell's principles and background, that his nation encouraged thousands of poor farmers to move to lands so dry that the settlers were bound to fail was a tragedy. He would spend most of the rest of his career trying to save them from that fate.

—

On 1 April 1878, in his role as director of the United States Geological Survey, Powell submitted a report to the commissioner of the General Land Office entitled *A Report on the Lands of the Arid Region of the United States, with a More Detailed Account of the Lands of Utah.*[4] Action on the report was urgent, Powell wrote, because "to the Arid Regions of the west thousands of persons are annually repairing." Scholars have been consistent in their judgment: "Quite possibly the most revolutionary document ever to tumble off the presses of the Government Printing Office";[5] a book that "did as much as any to lay out the fundamental choices facing the westward-moving democracy";[6] a report with "revolutionary implications even today";[7] "a complete revolution in the system of land surveys, land policy, land tenure, and farming methods in the West, and a denial of almost every cherished fantasy and myth associated with the Westward migration and the American dream of the Garden of the World."[8]

Powell's argument *was* revolutionary. Perhaps he had inherited a bit of the preacher from his father, though the son's devil was aridity, not Satan. Powell said that methods of agriculture that had worked elsewhere in America could not reclaim western lands. Nature had decreed that the West was different and there was only so much that man could do about it. To those whose beliefs, religious or professional, obligated man to have dominion over Nature, the Major's views were anathema. He pointed the way to a sustainable West, but to get there, politicians would have to prefer inconvenient facts to expedient fiction.

Arid Lands was only 195 pages long and the real meat was in the first two chapters, 45 pages of prose including charts, maps, tables, Latin names of species, and the like. From the very first page, the reader could see that Powell's opinions and recommendations would rest on a bed of scientific data. Those hoping to find evidence that rain follows the plow would have to look elsewhere.

Folded into a back pocket of the original edition and reproduced as the frontispiece of the 1983 Harvard Common Press edition was a remarkable map: the rain chart of the United States. Superimposed on the familiar state and territorial boundaries were a set of unfamiliar contours that marked out regions with different amounts of annual rainfall. The map depicted areas with more than twenty inches of rain a year, the minimum required for nonirrigated agriculture, in different shades of gray; those with less than twenty inches were shown in white. One could perceive the gist of the map from across a good-sized room. While the entire eastern half of the country was colored in one shade or another of gray, almost all the western half, save for a thin strip in the Pacific Northwest, was white. Serving as an approximate boundary between the two sections was the 100th meridian.

Here was undeniable, scientific proof that the country divided into wet and dry regions. Wallace Stegner summed up the import: "That one simple fact was to be, and is still to be, more fecund of social and economic and institutional change in the West than all the acts of all the Presidents and Congresses from the Louisiana Purchase to the present."[9] Stegner wrote in 1954; half a century later, his words are as true as ever.

Powell said that where ample rain fell, or where streams and lakes held enough water for irrigation, the 160 acres provided by the Homestead Act were more than a family needed or could farm successfully. He recommended that once government surveyors had certified land as having sufficient water for irrigation, a small group could form an irrigation district and divert, store, and use water. But no member of the district should have a parcel greater than eighty acres. Most significantly, Powell intended that "these lands are to be reserved for actual settlers, in small quantities, to provide homes for poor men, on the principle involved in the homestead laws." The members of the district would "establish their own rules and regulations for the use of the water and subdivision of the lands."[10]

After three years of irrigating his eighty acres, a farmer would gain title to both the land and the water. Attaching water rights to the land broke with the Eastern tradition of riparian law. In that venerable system, the person who owns the riverbank may use all the water he can take. But then the water must be returned to the stream. Riparian rights worked well when water drove a mill wheel or sluiced gold-bearing gravel. But an irrigator spread water over the land and returned little or none, leaving those downstream at his mercy.

Even if Powell were right about annual rainfall, surely farmers could take all the water they needed from western creeks and rivers. But Powell quashed that notion as well. Every drop of water in every western stream would irrigate only a small fraction of the land, he claimed. To illustrate, he calculated how much land all the streams in Utah could water. The total came to 2,262 square miles, 2.8 percent of the state's land area.[11] He may even have overestimated, since in 2003, Utah farmers irrigated

2.1 percent of the state's land area.[12] And that was after the Central Utah Project and other government irrigation programs brought water to the fields of the Beehive State.

At the time he wrote, Powell had witnessed how the Mormons treated water as communal property, organized themselves and cooperated toward a common goal, acceded to authority in water matters, and farmed on a small, family-sized scale—as he had done as a boy in Wisconsin. In the two decades the Mormons had been in Utah, they had demonstrated that the white man could survive in an arid land, but only by doing everything right. Underpinning Powell's proposals for the arid lands were the principles of Mormon irrigation.

Powell presented his recommendations in the form of two draft bills. His supporters in Congress duly submitted the pair, but both died in committee. The problem was that to accept the Major's argument a politician would also have to accept and admit that large sections of the West deserved that awful appellation: arid. Few elected officials from states west of the 100th meridian were willing to make that self-defeating admission.

Powell knew that a secure and productive system of dams and reservoirs had to be based on accurate topographic maps. How could he get the funding to prepare those maps in spite of the lack of support in Congress for his draft bills? By now an experienced Washington hand, the Major found a way. In his second budget request as director of the U.S. Geological Survey, he arranged for the insertion of a short phrase into a bill about "sundry civil construction": "to continue the preparation of a geological map of the United States."[13] After the congressman from Tennessee moved the amendment, these twelve seemingly innocuous words returned thousands of additional words of debate, a kind of congressional miracle of the loaves and fishes. States' rights was a major political issue at the time, and its proponents feared the new language was a trick by which the federal government would benefit at the expense of their states. They pointed out that since Congress had never authorized any such mapping, the survey could not lawfully "continue" it. But they were in the minority.

Though the legislators eventually approved Powell's language, most could not have imagined where he intended to take it. Anyone examining today the Geological Survey's magnificent geologic maps would have no trouble understanding the Major's plan. These works of science and art show not only rock formations, but also the underlying topography. As Powell would put it several years later, "geologic work could not be carried out without maps showing the relief of the land as well as the hydrography and culture."[14] But a topographic map that shows hydrography would also reveal the best places to site dams and reservoirs. Powell's clever insertion had opened the arid lands to his department's scrutiny. His purview, as he capaciously defined it, was everything from the 100th meridian west to the Pacific Ocean.

The Senate never overtly accepted the reforms that Powell outlined in *Arid Lands,* but it could not escape his chain of logic. By March 1888 one of Powell's scientific facts was undeniable: the West had too little water to irrigate all the land. To collect and best use what water did exist would require a system of dams and reservoirs. To discover where to site them would require an irrigation survey. To conduct the survey would require an experienced organization; the only candidate was the Geological Survey, already on the job "continuing" its preparation of a geological map. QED.

In order to prevent speculators from shadowing surveyors and buying up land, on 2 October 1888 the Senate amended the sundry civil appropriations bill to remove from settlement all lands that new reservoirs might irrigate. Here was revolution indeed. On a vast section of the United States, Congress closed public lands and suspended all public land laws. The test of the closure came within a year. Speculators in Idaho had filed claims on a prospective reservoir site, but the commissioner of the General Land Office, finding their claims incompatible with the legislation, canceled all claims made after the date the Senate acted. He even closed the land offices, leaving applicants with nowhere to file. The commissioner's ruling caused such an outcry from congressmen, speculators, and would-be settlers that his replacement submitted the matter for a ruling. The U.S. Attorney General upheld the decision; President Grover Cleveland agreed. Reopening the lands would have to wait for Major John Wesley Powell to report. How long would that take? No one, not even the Major, could say, but sitting politicians could see that it was likely to be well after they had gone to their reward.

For one government official to have more power than the president of the United States and more say-so over western states and territories than elected representatives was intolerable in general. It was intolerable in particular for that person to have single-handedly closed public lands to settlement, as his enemies accused Powell of doing, even though the blame lay with Congress itself. Powell, the bewhiskered bureaucrat, had gotten too big for his britches. Even the Major's friends and strongest supporters in Congress now deserted him. In 1890, the Senate removed the language that reserved the public lands, reopening them to settlement and speculation without benefit of maps. Powell's irrigation survey, like the man himself, was fast becoming irrelevant.

—

Venality and politics dissolved Powell's plan for western irrigation and delivered the most crushing defeat of his life. His was not the only dream of the day to dry up. In the late 1880s, a severe drought gripped not only the region west of the 100th meridian, but the subhumid wheat belt just to the east of it. The wet years that preceded the drought had drawn settlers west. Then rain failed to follow the plow or anything else: there was no rain. For many homesteaders, the only sane action was to get out and a large fraction did.

Did the drought discourage the irrigation boosters? Just the opposite. Then, as today, nothing better makes the case for new reservoirs than a long drought. Since Congress had yet to show any interest in having the federal government build dams and reservoirs, irrigation boosters decided to organize and lobby. The irrigation apostles founded an irrigation congress and at its second meeting, held in Los Angeles in October 1893, scheduled the Major to give two speeches. The delegation from the Geological Survey included his nephew, Arthur Powell Davis, who would go on to a prominent career in federal reclamation and become the proximate cause of Hoover Dam. The Major's first talk reprised his exciting adventure running the Colorado River through the Grand Canyon more than two decades earlier. Railroad engineer Robert Stanton scolded Powell, charging that the Major had overlooked the wealth potential of the West, especially its canyons. They should become the sites of dams, dynamos, and even railroads, said Stanton. The assembled delegates urged the federal government to build dams and reservoirs to provide water to innumerable easterners striving to wrest themselves from a landless condition. As Worster observes, this was a self-serving analysis of what poor easterners wanted, none of whom were in Los Angeles to speak for themselves.[15]

By the time Powell rose a second time, he had spent three days listening to delegates describe how irrigation could reclaim far more of the West than he knew was possible. It must have seemed as though *Arid Lands* had never been written, as though his twenty-year crusade had amounted to naught. Not yet sixty, afflicted by his injury, his enemies in Congress, his shrewish wife, and now by the heresy to which he had been subjected, Powell may well have seemed to be the "old man" that he called himself. We might imagine him as an aged, nearly worn-out bull buffalo, standing on a high bluff from which he could see the West spread before him, not liking what he saw. But he was not done yet.

Powell told the attendees that he was not going to deliver his prepared technical talk on water supplies, but instead would address the important problems that had come before the congress. The delegates, unaware of what the Major had in mind, applauded. When he said, "I am more interested in the home and the cradle than in the bank counter," they applauded again. But that was the last time they did so during these remarks.

Powell complemented the report before the congress, then proceeded to demolish its underlying assumption:

> When all the rivers are used, when all the creeks in the ravines, when all the brooks, when all the springs are used, when all the reservoirs along the streams are used, when all the canyon waters are taken up, when all the artesian waters are taken up, when all the wells are sunk or dug that can be dug in all this arid region, there is still not sufficient water to irrigate all this arid region.[16]

It was merely simple arithmetic, he said. If you need twenty-four inches of water a year to grow crops on an acre-foot of land, and Nature supplies "three, four, or five inches of rainfall," even if you catch every drop it will not be nearly enough.

Possibly lending unintended credence to the false accusation that he had caused the closing of the public lands, Powell insisted that "not one more acre of land should be granted to individuals for irrigating purposes." At an irrigation congress, here was heresy indeed. The delegates leapt to their feet, booed, and shouted the Major down. The representative from Mexico opined that "it was the only bullfight he had yet seen in this country."[17] In response, Powell retorted, "What matters it whether I am popular or unpopular. I tell you, gentlemen, you are piling up a heritage of conflict and litigation over water rights, for there is not sufficient water to supply these lands." A few months later, Powell laid out his facts and statistics in a magazine called *The Irrigation Age*. But the true believers, whose faith could trump any number of scientific facts, were unpersuaded.

Soon his enemies in Congress, who had grown more numerous, forced Powell to resign from the Geological Survey, leaving him with only his sinecure at the Bureau of Ethnology. There the Major waited out his days. As Stegner put it, "It was the West itself that beat him, the Big Bill Stewarts and Gideon Moodys, the land and cattle and water barons, the plain homesteaders, the locally patriotic, the ambitious, the venal, the acquisitive, the myth-bound West which insisted on running into the future like a street-car on a gravel road."[18]

—

In late May 1889, heavy rainstorms moved into western Pennsylvania, swelling the creeks and the Conemaugh River. Above Johnstown sat the privately constructed South Fork Dam, at one time the largest in the world. Since its completion in 1853, the dam had passed through several owners and several failures. Luckily, because the reservoir was only half full, an 1862 collapse had done little damage. But in mid-afternoon on 31 May, the dam liquefied, sending a thirty-foot wall of water at speeds of up to forty miles per hour down on Johnstown. The flood washed away the northern half of the town and killed over 2,200 people, more than the San Francisco earthquake seventeen years later. Until the World Trade Center attack in 2001, only one disaster in United States history had taken more lives: the Galveston Hurricane of 1900, in which 6,000 perished. The Johnstown Flood was especially tragic because authorities could never identify hundreds of bodies.

Reaction was swift. Many indicted privately built dams, whose owners did not have to meet government safety standards; others were ready to give up on dams altogether. Powell believed that to be a serious overreaction. The Johnstown dam "was properly constructed," Powell wrote, "[but] the works were not properly related to the natural conditions." Before building a reservoir, one should "determine the amount of water to be controlled."[19] That requires topographic and hydrographic surveys. "Facts," he wrote, "are to be collected as preliminary to the construction of a reservoir system. To neglect the essential facts is to be guilty of criminal neglect." Properly planned, sited, and constructed dams and reservoirs, using the principles he had outlined in *Arid Lands,* could lower the risk of flood damage to an acceptable level.[20]

By the middle of the 1890s the vision of an irrigated, prosperous West was in trouble. The Homestead Act could hardly be deemed a success: the lack of water had caused too many settlers to fail. The Desert Land Act, the Swampland Act, and their ilk had done more for speculators and outright crooks than for honest yeomen. Private irrigation was demonstrably dangerous. Dams that held up physically were apt to fail financially.

Since the federal government showed no inclination to take on large-scale irrigation, several states elected to do so themselves. The Wright Act of 1887 established irrigation districts in California, but they had trouble selling their bonds and, at times, filling their reservoirs. When a reservoir did fill, district officials often could not decide how to allocate the water. Ten years later, the California legislature gave up. Colorado tried a similar program, adding a modest financial sweetener. Five dams resulted, but none were successful. The senator from Wyoming decided to enlist the federal government, which would provide up to one million acres to any state that would see that the acreage was irrigated. The states were to sell the land to farmers, with the federal government receiving the sale proceeds. By 1902, only four states had even applied for the funds and the land irrigated amounted to less acreage than a typical western county.

A logical response to these failed efforts would have been to give up on federal reclamation. Nothing had worked. Oh, ye of little faith! Each failure led the boosters to redouble their efforts to rope in the federal government, whose deep pockets were needed to fund their schemes. Federal dam construction need not drain the treasury, they said; once the dams and irrigation works were in place, farmers would purchase water at a fair price and pay the government back over time. The small farmer did not need a handout—just a helping hand. Various protections would prevent speculators, capitalists, and big business from monopolizing the land. As we will see, none of these claims came true.

In late January 1901, Francis Newlands of Nevada introduced a bill in Congress that encapsulated the dreams of the boosters and showed in detail how to fund federal reclamation. His bill would establish a fund that would pay for the irrigation works initially. Farmers would buy irrigable land and pay the reclamation fund back in ten annual installments, interest-free. Once the fund reached steady state, it would finance irrigation projects indefinitely.

Newlands had tried private irrigation and lost half a million dollars, a vast sum in his day, though not enough to break a man married to a daughter of the Comstock mining clan. The experience convinced him that only the federal government was up to the task of large-scale reclamation. Newlands found an important ally in Theodore Roosevelt, who had become president on 14 September 1901 after an assassin's bullet felled McKinley. Echoing the claims of the boosters, Roosevelt announced that "the western half of the United States would sustain a population greater than that of our whole country today if the waters that now run to waste were saved and used for irrigation."[21]

But many remained opposed. Eastern congressmen protested that Newlands's bill would provide unfair competition to the farmers in their states. Why should western farmers get a government handout—which is what the program would amount to when it turned out that farmers could not repay their loans—while easterners did not? Crops were already in surplus.

The national taxpayer would foot the bill, but only a few would benefit. And, echoing an argument made in response to the 1882 amendment to continue preparation of a geologic map, "Federal reclamation violated State rights and was therefore unconstitutional."[22]

But these arguments carried little weight. The congressional members debated now in a new century, yet two-fifths of the country remained too dry to join in the nation's prosperity. With the president strongly in support, on 17 June 1902, Congress passed the Newlands Act and launched the era of federal reclamation. Instead of Powell's recommended eighty acres, Congress set the maximum at 160 acres. But at least the legislators had set *some* limit. This would prevent capitalists and speculators from assembling parcels of thousands of acres. Since the act required settlers to be "actual bona fide resident[s] on such land," federal reclamation would serve Powell's poor men.

A question that has intrigued every student of John Wesley Powell is what he would have thought of the mega-dams and reservoirs that tame the wild river of his adventures. Would he have been proud to have one of the giant reservoirs bear his name? Powell may have given a clue when he predicted that, because there was enough water to irrigate only a small percentage of the West, "all the waters of all the arid lands will eventually be taken from their natural channels."[23] From this passage, master historian and Powell biographer Donald Worster surmises that "even the mighty Colorado, running through its stupendous canyons, a river that had afforded Powell the most thrilling moments of his life, must someday become a dried-up channel as irrigators diverted its entire current to their fields."[24] Indeed, before the river reaches the Gulf of California today, it has shrunk to a trickle, or more often, disappeared altogether. But Powell wanted big government and big business to stay out of western dam-building. Consistent with the plans he outlined in *Arid Lands,* his vision for the Colorado River would have been for groups of local farmers, who lived on their land, to build small reservoirs on the accessible tributaries of the Colorado River, high above the abyssal canyons.

Powell may have foreseen the end of the Colorado River, but he could not have foreseen the means. To conceive of Hoover and Glen Canyon dams and Lakes Mead and Powell, the Major would have needed more imagination than his contemporary, the visionary writer Jules Verne. When Powell wrote *Arid Lands* in the 1870s, the concrete arch dam had not been invented; the first in the United States did not rise until 1903, one year after his death. No construction company of his day would have had any idea how to build a giant dam at the bottom of a deep, hard-rock canyon in a roadless, wild region. Had such a dam been built, how would the water have been lifted

Figure 5. River's end. The Colorado River suffers not one death but several. It first sinks into the sand and dies downstream of Morelos Dam near the border. Agricultural runoff brings the river back to life and sends it on to the Rio Hardy, after which it disappears once more. This scene shows the river's final demise. (Courtesy of Karl Flessa)

to the rim? Not by electrically powered pumps, for no hydropower dam had yet been built in the United States.

Powell could not have foreseen that giant concrete dams larger than the pyramids would impound reservoirs nearly two hundred miles long. That hydropower dams would generate the electricity to pump water thousands of feet vertically and send it hundreds of miles laterally through complex networks of pumps, pipelines, canals, and siphons. That the landowners who benefited would not be the local farmer and rancher, but giant agribusinesses and absentee corporations who owned thousands and tens of thousands of acres. That the principal crop on federally irrigated lands would be food for someone else's cattle. The hydraulic society of the American West became and remains the opposite of everything for which John Wesley Powell, Thomas Jefferson, and Francis Newlands stood.

FIVE

The Reality of Empire

WITH THE DREAM OF A federal reclamation program about to come true, even the most fervent irrigation booster might have been excused for hiding a few private doubts. If irrigation were the panacea, why had so many private and state schemes failed? Would enough land-poor eastern-ers take up the government's offer of a loan and head west? Would they be able to repay their government loans and make the reclamation fund self-sustaining? What if another drought like that of 1887–90 descended on the West? Federal reclamation had many ways to fail and did not even have to fail completely. A fickle Congress, fearing that the program would become an embarrassment or deciding that it had a greater need for the money elsewhere, might withdraw funding at any time. The new Reclamation Service needed successes, and quickly.

The program could have started with a few, carefully chosen, small-scale projects. This would have allowed Reclamation Service engineers to put their theories into practice and gain valuable experience before scaling up to larger dams. Or, the service could have gone for broke and built the largest single dam it could, setting a spectacular example and drawing a favorable contrast with the smaller and often unsafe private dams. Instead, more for political than engineering reasons, the service soon had thirty projects going, enough to guarantee that it would be unable to learn from deliberate experience.[1]

One of the projects, on Arizona's Salt River east of present-day Phoenix, looked to be ideal for producing immediate results. The earliest white settlers there had begun by excavating and re-using Hohokam canals. When they needed more water, private companies built new irrigation works. By 1900, nearly 20,000 valley farmers cultivated 130,000 acres, the maximum the Salt River could support without a storage reservoir upstream to control flooding and supply water during dry periods.[2] Valley residents had joined national reclamation enthusiasts to lobby on behalf of Newlands's bill. Now they successfully petitioned their representatives to make a dam on the Salt River one of the first federal reclamation projects.

The dam site was so remote that to get to it workers had to cut a road, called the Apache trail, through sixty miles of the rugged Superstition Mountains of eastern Arizona. To build the dam, stonecutters and masons, including some Apaches, carved the local stone into six- and ten-foot blocks and joined them together. When completed in 1911, Theodore Roosevelt Dam was a beautiful stone masonry construction, curving gracefully athwart the Salt River, looking like something the Romans might have built. And well they might have, since the Romans also built arch dams out of stone blocks.

The Newlands Act intended that farmers would settle and reclaim arid *public* lands. But just the opposite happened. Locating the Salt River project where people were already farming obviated the need to entice them there, but it also meant that the project would include not a single square foot of public land. That turned out to be the rule. By 1910, fewer than half the projects the Reclamation Service had under way were on public land.[3] Of course, a settler could always buy private land. But the more irrigable the acreage, the higher the price. Thus in its first substantial project, the Reclamation Service served no poor easterners and reclaimed no public land. What the Salt River project did do was establish the precedent that federal reclamation could ignore the intent of Congress and get away with it.

Poor people would have had a better opportunity to acquire land had another stipulation of the Newlands Act been followed. The legislation said that "public lands which it is proposed to irrigate . . . shall be subject to entry only . . . in tracts of not less than forty nor more than one hundred and sixty acres."[4] That language should have prevented speculators and capitalists from assembling large parcels, leaving more land for small farmers. President Theodore Roosevelt described the national policy in 1906, four years into the federal program: "When a man attempts to hold 160 acres of land completely irrigated by Government work, he is preventing others from acquiring a home. Speculation in lands reclaimed . . . must be checked at whatever cost. The object of the Reclamation Act is not to make money, but to make homes."[5]

Large landowners who wanted federal water would simply have to sell enough land to get below the 160-acre limit. "Not one dollar will be invested," proclaimed Frederick Newell, director of the Reclamation Service, "until the Government has a guarantee that these large farms will ultimately be put in the hands of small owners, who will live upon and cultivate them."[6] At the irrigation congress twelve years before, Major Powell had been equally blunt. Irrigation disciples had shouted him down and in the intervening years had gotten everything they wanted. Why should they now pay attention to Newell, the latest irritating federal bureaucrat? Ignore him, as they had ignored Powell, and before long powerful men would see that Newell's warnings carried no more weight than had Powell's.

Indeed, the first time federal reclamation had an opportunity to enforce the acreage limit—in the Salt River project—it balked. The failure was so blatant and pervasive that a 1916 independent review said it "closely verges on fraud."[7] Instead of enforcing the limit, in that same year the service raised it. The Newlands Act had specified that the 160-acre limit applied to "any one landowner." Now the Reclamation Service ruled that a husband and wife would each be landowners, allowing a couple to own 320 acres. In the coming decades, the acreage limit would rise repeatedly—and each time the government would fail to enforce it.

Not only could few settlers afford to buy private land, the hordes who were supposed to head west and take out a government loan on public land mostly failed to show up. The relative few who did tended to be much less prepared than those who had come west a generation or two earlier. As Michael Robinson put it in his quasi-official history of the Bureau of Reclamation (as the service was renamed in 1923): "Western economic and social determinants were changing rapidly. Nineteenth-century irrigation pioneers were better suited to endure hardships than settlers who struggled to survive on Federal Reclamation projects after 1902. In the nineteenth century, wild game was plentiful, livestock could graze on the public domain outside irrigated areas, and the settlers were inured to privation." Those earlier pioneers also knew enough to settle on land that already had water. The later arrivals knew nothing about irrigation—but then neither did the Reclamation Service. Robinson again: "Initially, little consideration was given to the hard realities of irrigated agriculture. Neither aid nor direction was given to settlers in carrying out the difficult and costly work of clearing and leveling the land, digging irrigation ditches, building roads and houses, and transporting crops to remote markets." Without knowledge or assistance, settlers flooded and waterlogged their lands and allowed silt to clog their irrigation systems. They filed on more acres than they could irrigate, leaving them with more debt but no more income.[8]

Robinson explained, "There was a tendency for some engineers to view public works as ends in themselves. Despite official declarations from more sensitive administrators that 'Reclamation is measured not in engineering units but in homes and agricultural values' . . . the service regarded itself as an 'engineering outfit.'"[9] Engineers knew how to design and build dams, but in the early 1900s, no one knew how to use social engineering to obtain a desired result. Reclamation Service engineers understandably stuck to what they did know and left the social consequences to others, or more often, to no one. A modern review found that a century later, little had changed. In 2006 the National Research Council reported that "reclamation employees appear on the whole to be more motivated by complex technical tasks than by tasks that are socially and politically complex."[10]

Powell had outlined a careful approach that would have sited reclamation projects where topographic maps showed they would store the most water at the least cost. Instead, politics and pork ruled. "The government was immediately flooded with requests for project investigations," wrote Robinson. "Local chambers of commerce, real estate agents, and congressmen were convinced their areas were ideal for reclamation development. State legislators and officials joined the chorus of promoters seeking reclamation projects . . . legislative requirements and political pressures sometimes precluded careful, exhaustive surveys of proposed projects . . . projects were frequently undertaken with only a sketchy understanding of the area's climate, growing season, soil productivity, and market conditions."[11] Many projects thus wound up where they were bound to fail, though no one told the homesteaders.

The federal reclamation program soon got into financial trouble as well. As farmers paid back their interest-free loans, the reclamation fund was supposed to self-perpetuate. But even with ten years to repay, most farmers could not. By 1910, the fund had paid out so much more than it took in that it verged on bankruptcy. Congress stepped in with a $20 million loan and, when that proved insufficient, doubled the loan repayment period to twenty years. It also allowed the fund to sell surplus water to farmers outside an irrigation district. In its report to the secretary of the Interior, a review commission noted that twenty-one years into the program of federal reclamation, water users had been able to repay less than 11 percent of their obligation.[12]

Western farmers were demonstrably worse off than their eastern counterparts. Most had gambled everything, including the health and stability of their families, on the promises of irrigation boosters, reclamation officials, and politicians. They worked themselves into the ground in a futile effort to make a living on land that for ten thousand years had supported only nomadic bands. As Donald Worster eloquently summed up their plight: "This, then, was the reality of empire: the family fighting to hold on to a forty-acre place in the sun, grubbing out sagebrush and cactus to raise a redundant crop like wheat, stuffing rags into the cracks to stop the winter wind moaning through a bare-board shack, all for so little income, so uncertain a future."[13]

Though many small farmers had gone under, perhaps the program still served the nation by providing vital crops that could not be grown as well elsewhere. Not so. Three-fourths of reclamation project acres were planted in crops already in surplus: wheat, cotton, alfalfa, and other pasture grasses. Well then, perhaps federal reclamation would find its redemption in the farmers who stuck it out and eventually came to own the land they farmed. After all, the Newlands Act had specified that a reclamation landowner must "be an actual bona fide resident on such land, or occupant thereof residing in the neighborhood of said land." But like the acreage limit, the residency requirement was honored in the breach. When farmers failed, absentee landowners bought them out and hired tenant farmers. "Actors, artists, teachers, undertakers, and railroad workers," none of whom could remotely be considered resident farmers, complained to the bureau about their ineffective tenants.[14] In a 1916 textbook, Reclamation Service director Newell blamed the victim.

> The irrigators as a body are not only inexperienced, but many of them are disappointed in that they have expected easier things. Thus they do not always appreciate the efforts made in their behalf. There has been attracted to the locality a considerable number of men who have never made a success elsewhere; these attribute their failure to make good under the new conditions not to their own inability, but largely to the faults of the country or of the system.

There has thus arisen a class which has been called the "professional pioneer," always seeking for something a little better or for conditions where life will be easier; staying in any one locality only a few months or years and then again seeking El Dorado.[15]

Others had already decided the fault lay with Newell. In 1915, the secretary of the Interior replaced him with Arthur Powell Davis, the Major's nephew, who by now had earned excellent credentials. He had served as topographer with the Geological Survey and led the study of the route of the Panama Canal. One of the first to join the new Reclamation Service, by 1906 he had become its chief engineer. Davis stayed with the service until 1923, rising to director and leaving just as the agency became the Bureau of Reclamation.

Federal reclamation had not merely fallen a bit short of the promises of the irrigation boosters—it had failed in nearly every conceivable way. An objective analysis might have concluded that it was time to shut the program down and spend the taxpayer's money on something more productive. But if there is one clear lesson from the history of reclamation, it is that once begun, individual projects, much less an entire program, are seldom if ever shut down.

Nevertheless, something did have to be done to avoid embarrassment. The answer, then as now, was to appoint a commission. Appointed on 8 September 1923, the seven members of the Committee of Special Advisors on Reclamation were John Widtsoe, a university president and irrigation expert; James R. Garfield, Theodore Roosevelt's Interior secretary; plus a former governor of Arizona, the present commissioner of Reclamation as well as the man who would succeed him the following year, the president of the United States Chamber of Commerce, and the president of the American Farm Bureau. These men were capable of criticizing the performance of the Reclamation Service; indeed they reported that "unless remedial measures of a permanent character are applied, several more projects [than the three that had failed] will fail; and the Federal reclamation experiment conceived in a spirit of wise and lofty statesmanship, will become discredited."[16] But not one of the committee's sixty-six recommendations raised serious question about the need for a program of federal reclamation, regarded as both a God-directed obligation and a practical imperative. On the committee's advice, Congress extended the loan repayment period to forty years, allowing the subsidy to approach or exceed the sum paid for the land in the first place.

By the time the Committee of Special Advisors reported in 1924, large surpluses had dropped crop prices so low that farmers could neither make their loan payments nor pay for their water. By 1927, rising land prices had caused one-third of reclamation farmers to sell out. As the 1930s began, few reclamation projects had paid for themselves; farmers had voted out with their feet; by far the majority of irrigated acreage was on private land; absentee landowners and speculators owned large tracts. Nearly thirty years after the Newlands Act, federal reclamation had betrayed every one of its founding principles.

In a 1930 publication, former Reclamation Service director Newell, as knowledgeable a person on the subject of water in America as could be found, condemned the program in which he had invested so much. "There [had not been] a real need for more farm land," he wrote. "The demand which seemed to exist for irrigated areas was largely artificial, stimulated by the speculative spirit; it did not come from the true economic needs of the country."[17] Newell said that the program of federal irrigation had raised total crop production by less than 1 percent, yet Congress had spent more time and energy on that minute fraction than it had on the other 99 percent. Reclaimed lands had often proved to be worth less than the money it took to irrigate them. Because of the program, Newell contended, "There has come about a disastrous loss in the higher ideals of citizenship, and that the invaluable qualities of self-reliance have been undermined by the insidious forms of paternalism, more than offsetting the crop gains."[18] According to Newell, to achieve such meager results the national taxpayer would have to fork over four times as many dollars per acre as the farmer who owned the land.

Everything that could resuscitate federal reclamation had been tried. Collectively the various band-aids might keep the program alive, but they could not guarantee good health and prosperity. If the program limped along, who could say that some future Congress or president, conducting triage, might not eliminate the Bureau of Reclamation? The agency responded with one of the options it had avoided early on. It would build a dam so large as to be a wonder of the modern world, thus forever establishing the value and necessity of federal reclamation.

⌒

A dam and reservoir of the size the bureau had in mind would have to meet several criteria. First, to maximize flood control and supply water during droughts, the reservoir needed to hold at least twice the annual flow of the river. The dam should generate enough electrical power to pay for its construction, quieting those who might object to its cost. The deeper the reservoir, the greater the hydraulic head on the turbines and the more power the dam would generate. A deep reservoir also has a smaller surface area relative to its volume and so evaporates a smaller fraction of water than one that spreads out over the countryside. The site should be close to the fields that the water would irrigate. Finally, the fewer people who lived nearby, the fewer to object when the reservoir inundated the land. These criteria all pointed to a site that was narrow, deep, and long: a canyon. The chasm should be remote but accessible.

But a dam site in a deep canyon also has disadvantages. First, to carry men and equipment down to the level of the river and supply them would be difficult, costly, and dangerous. Second, contractors would have to emulate Moses and stop a wild river dead in its tracks so that the dam site could remain dry during construction. The only feasible way to do so was to block the river with a cofferdam and divert the water through bypass tunnels cut through the bedrock walls. These passages might have to be a thousand feet long or more, adding to the timetable, the cost, and the danger. Third, water in a deep, narrow reservoir exerts great pressure on a dam. The only type strong enough to withstand the stress is a dam that curves upstream—like Roosevelt Dam—in the shape of an arch. That form puts a dam under compression, wedging the structure against itself and making it stronger. But the large stone blocks used at Roosevelt Dam would be impossible to cut and manipulate at the bottom of a deep canyon. In 1910, in the canyon of the Shoshone River in Wyoming, the Reclamation Service had experimented with a dam made of poured concrete. At 325 feet high, Buffalo Bill Dam was the first to be taller than wide. Mules had hauled the cement in wagons seven miles to the dam site west of Cody, where it was dumped into forms. The method worked, but unless the bureau wanted to hire an army of muleskinners, it could hardly be scaled up to the giant dam the Bureau of Reclamation now planned. Between 1911 and 1915, at Arrowrock Dam in Idaho, the service had learned how to build a larger dam by mixing concrete at the site and pouring it directly into frames, avoiding the obstinate mules and allowing even larger dams.

For a river to flow in a narrow canyon takes a special set of geological circumstances. The Grand Canyon may be the ultimate expression of a western canyon, but the Colorado Plateau has experienced sufficient tectonic uplift and subsequent erosion to have many others. Several sites on the Colorado River would permit a dam six or seven hundred feet high that could hold back twenty-five to thirty million acre-feet of water, twice the river's annual flow.

What attracted the bureau to the Colorado was not that it was the longest river in the country—the Missouri, Mississippi, Rio Grande, and St. Lawrence are each lengthier. Nor does the Colorado River carry the most water: by volume, it ranks only about 25th in the country. The attraction of the Colorado River is its steep fall: 11,000 feet in 1,450 miles, more than ten times the gradient of the vastly larger Mississippi. The Colorado's sharp descent had cut the most spectacular canyons on the continent; now it could spin turbines faster than any other large river in the country.

John Widtsoe gave another reason for damming the Colorado River: "The destiny of man is to possess the whole earth; and the destiny of the earth is to be subject to man. There can be no full conquest of the earth, and no real satisfaction to humanity, if large portions of the earth remain beyond his highest control. Only as all parts . . . are . . . brought under human control can man be said to possess the earth."[19] To bring one large portion of the country—the Southwest—under control, man had to conquer the Colorado River. That was not a choice but a sacred obligation.

Engineers had their own reasons for wanting to tame the Colorado: the bigger the dam and the wilder the river, the greater the challenge and the greater the accomplishment. Early in his career, A. P. Davis spoke for the dam-builders: what the West needed, he said, was "the gradual comprehensive development of the Colorado River by a series of large storage reservoirs," starting with a dam beyond the mouth of the Grand Canyon that would be "as high as appears practicable from the local conditions." Later Davis admitted that no project in the West had "excited [his] interest and imagination and ambition so much as the development of the Colorado River basin."[20]

In the early days of federal reclamation, the agency did not have to show that the economic benefits of a project exceeded its costs. But it did have to explain how the water of the planned reservoir would irrigate fields and bring new land into production. The terrain through which the Colorado flowed made this a daunting task. For most of its length, the river traverses country so high and rugged that above the canyon rims there are no fields and pastures to irrigate. Beyond the mouth of the Grand Canyon, though, the river verges on a potentially fertile region of low relief. But the tract was so deadly dry that people called it the Colorado Desert. Immediately upstream lie deep canyons that could hold the high dam that Davis and the bureau envisaged. If the agency could build the greatest dam in history and reclaim the most Godforsaken desert in the country, the Bureau of Reclamation would have done more than restore its reputation. No place in the West would be off limits.

SIX

This Vast Plain of Opulent Soil

IN THE YEAR 1774, THE VICEROY of New Spain ordered Captain Juan Bautista de Anza to travel west from the Spanish settlement of Tubac, near present-day Tucson, in search of a route to the missions on California's coast. The party forded the Colorado near its junction with the Gila, one of the most desolate and potentially fatal regions on the continent, and eventually reached Mission San Gabriel in today's East Los Angeles. Afterward, Anza said that the trip had been "*la jornada de los muertos.*"[1]

In 1883, William P. Blake, a graduate of the Yale Scientific School, surveyed the area near the confluence of the two rivers, which by then had become known as the Colorado Desert, for a railroad route. The young geologist recognized the mounds of travertine and wave-cut benches that lined the surrounding hills as evidence of a now-vanished lake. Only a few centuries before Anza and his men had blazed a trail under a deadly sun, Indians had splashed and sailed. But from where had a lake in the middle of a desert come and where had it gone? The answer is that the Colorado River, debouching from the Grand Canyon, had dropped so much silt as it slowed that it dammed its own path and impounded ephemeral lakes. Meandering back and forth behind these self-made obstructions, the river would first spill over into the Colorado Desert; when silt blocked that path, like a giant snake swinging its head from side to side, the river would change course and spill into Sea of Cortés. Lakes grew, shrank, and disappeared, only to reappear as the process repeated. The most recent example, called Lake Cahuilla after a local Indian tribe, was 100 miles long, 35 miles wide, and 300 feet deep, about the size of today's Lakes Mead and Powell.

In 1892, the Colorado Development Company sent C. R. Rockwood to Yuma to study how to divert the Colorado River to irrigate a section of Mexico's Sonoran Desert.[2] Rockwood soon saw that turning the water in the other direction, north toward the Colorado Desert, made more sense. He came to believe that by simply cutting a notch in the riverbank, water would flow downhill and irrigate as much as two million acres in the United States. Rockwood hired George Chaffey, who had laid out the town of Ontario, California, to work out the details. Chaffey evidently had what today we call marketing skills: one of his first acts was to rename the expanse of the Colorado Desert from today's Salton Sea to the Mexican border the Imperial Valley.

National irrigation figures, recognizing that transforming a desert into an empire would prove the value of irrigation, joined in the effort. The leader of the national irrigation boosters was William Smythe, who as a young man had moved from Massachusetts to Nebraska and become a newspaper editor. Living as he did near the 100th meridian, Smythe saw for himself how a lack of water could dash the dreams of even the hardest-working family. He took up the cause of irrigation, writing a book called *The Conquest of Arid America* and founding the magazine in which Major Powell published his heretical conclusions. In a 1900 travelogue in *Sunset Magazine,* Smythe waxed eloquent about the opportunity:

> This vast plain of opulent soil—the mighty delta of a mighty river—is rich in the potentialities of production beyond any land in our country which has ever known the plow. Yet here it has slept for ages, dormant, useless, silent. It has stood barred and padlocked against the approach of mankind. What is the key that will unlock the door to modern enterprise and human genius? It is the Rio Colorado. Whoever shall control the right to divert these turbid waters will be the master of this empire. Without the right and the ability to use this water nothing is possible; with it, everything is possible. In no part of the wide world is there a place where Nature has provided so perfectly for a stupendous achievement by means of irrigation.[3]

Chaffey began dredging in August 1900 and soon had a canal running from the Colorado River to an old channel called the Alamo River. From there the water would travel forty miles due west to a point south of the Imperial Valley and, passing just east of Calexico, flow north across the border.

On 14 May 1901, the first water from the Colorado River reached Imperial Valley fields; settlers filing under the Desert Land Act soon followed.

At first it appeared that optimism, the right name, and sound engineering could indeed transform an awful desert into a valley of empire. But those who thought so had not reckoned with the inexorable tendency of a river to keep on doing what it has always done. For the last five million years, the lower Colorado River has carried an average of more than a hundred million tons of silt each year. Now the silt clogged Chaffey's canals, requiring his company, like the Sumerians, to spend increasing time and money cleaning the irrigation works. By the summer of 1904, the river had won. Rockwood decided to start afresh by making a new cut, this time across the border, in Mexico. In 1905, before the Mexican government had approved a permanent headgate to control the flow of water through the new opening, the Colorado began to flood. The water poured through the cut and rushed west and north, sweeping the loose sediment before it. Soon the original notch was a mile wide, allowing the entire flow of the river to surge though the gap and flood Imperial Valley fields. Close to where the waves of Lake Cahuilla had once lapped, a new lake, the Salton Sea, rose almost overnight to become the largest inland lake in the country.

Nearly broke by now, the California Development Company had no hope of plugging the giant breach. In desperation, the company appealed for help to the Southern Pacific Railroad and its owner, magnate E. H. Harriman. The railroad baron had a stake in the crisis, for the company's rail lines ran right across the flooded valley. The shrewd Harriman agreed to try to seal the opening, but only on condition that he take over the development company. Railroad car after railroad car lumbered up to the edge of the cut, dumped its load of rock and gravel, only to have the river wash the debris away. This went on for two years. Finally, in February 1907, the railroad got the better of the river, pushing the Colorado back into its original channel. Twelve thousand acres of cropland had been lost and many farmers had been forced to sell. Those who stayed knew that more floods would drive them from the Imperial Valley for good. As if to confirm their fears, in 1910 the river flooded the valley again. This time, the Southern Pacific declined to help, requiring Congress to step in with a $1 million bailout. Anyone could see that next time, it might be Congress who said no. Then the valley of empire might return to a *valle del muerto*.

The wild river was not the only threat to the Imperial Valley. In return for hosting the canal intake, the Mexican government demanded up to half the water extracted from the river, with its share coming out first. Mexican consumption would not amount to much at the start, but Imperial Valley farmers feared that when Mexico began to take its full share, there might not be enough water for those north of the border.[4] To make matters worse, the Mexicans refused to help pay for the irrigation works. The last straw for valley residents was the revelation that most of the irrigated land in Mexico belonged not to Mexicans but to an American syndicate. The group included Harry Chandler, son-in-law of Harrison Gray Otis, whom he would succeed as publisher of the *Los Angeles Times*. The syndicate had shown no compunction about surreptitiously stealing water from Americans in the Owens Valley, hundreds of miles away on the west flank of the Sierras, and piping it to the San Fernando Valley. Men with so great a thirst but so few scruples were surely capable of pilfering Mexican water.

The farmers of the Imperial Valley needed the protection of an intake and canal system located entirely within the United States—an "All-American" canal. In 1911, to lobby for the canal and to provide an organization to manage it and the associated irrigation works, landowners established the Imperial Irrigation District. For several years the district held out the hope that it could build the canal without federal assistance, but only the federal government had pockets deep enough. As long as another devastating deluge remained a possibility, banks might refuse to lend to the absentee landowners who were assembling large parcels in the Imperial Valley and insurance companies might decline to insure them. A. P. Davis, by now director of the Reclamation Service, pointed out that by itself the canal would do nothing to prevent floods, whereas with "complete storage the flood menace would be removed." Davis saw "the Imperial Valley problem . . . [as] inseparably linked with the problem of water storage in the Colorado Basin as a whole."[5]

Davis ordered the Reclamation Service to reconnoiter Boulder and Black canyons, two chasms through hard igneous rock some thirty to forty miles southeast of a dusty Mormon backwater called Las Vegas ("The Meadows"). The service needed no encouragement: one of its first reports, back in 1902, had already described reconnaissance investigations of the two canyons. In 1920, based on new reports, Congress directed Davis to study the possibility of a high dam on the lower river. He and his co-author, secretary of the Interior Albert Fall, recommended a dam at Boulder Canyon and an All-American canal. With the two of them in agreement, one would have thought that the location of the dam would have been settled. But that did not reckon with Eugene Clyde LaRue, an obstinate hydrographer who knew more about the Colorado River than anyone and who believed that he knew of a better dam site, hundreds of miles above Black and Boulder canyons.

LaRue joined the Geological Survey in 1904 and spent years roaming the Colorado Plateau and floating the Colorado River and its tributaries, exploring for the best dam sites. In a 1916 report, he repeated John Wesley Powell's unwelcome message that the Colorado contains too little water to irrigate all the land in its basin. Given the scarcity, LaRue's principal criterion for siting reservoirs was to minimize evaporation. A dam at Boulder Canyon would impound a lake with a vast surface area in a region where temperatures often reached 120 degrees, both factors that increase evaporation. Dams farther upstream, where canyons were narrower and temperatures lower, would save water. LaRue went on to describe a series of thirteen small dams between the mouth of the Grand Canyon and Cataract Canyon, hundreds of miles upstream just below Moab, Utah, a plan that would turn the wild river into an echelon of millponds. The linchpin of LaRue's scheme was a five-hundred-foot dam at a remote place called Glen Canyon.

Compared to a single dam beyond the mouth of the Grand Canyon, LaRue's multi-dam plan would store more water, reduce surface area and evaporation, increase power generation, and regulate it better. His scheme

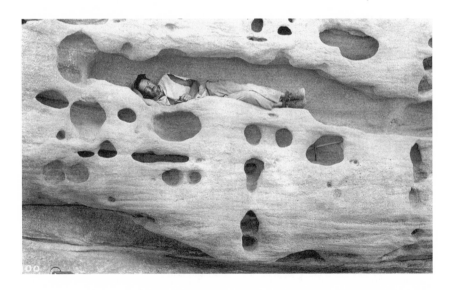

Figure 6. Hydrographer E. C. LaRue resting in the Grand Canyon one mile below the mouth of the Little Colorado, 1923 (U.S. Geological Survey)

made engineering sense. But to build thirteen dams in some of the worst terrain in the nation would be expensive and take so long that it would extend well beyond the career of any sitting politician or bureaucrat. Moreover, the Bureau of Reclamation by now had lost considerable credibility. Before endorsing such a complex and far-reaching program, Congress would likely have needed convincing that federal reclamation deserved a future.

—

While LaRue surveyed the Colorado River, in 1922 the United States Supreme Court handed down a ruling that struck fear into each Colorado River basin state other than California. The issue in *Wyoming v. Colorado* concerned water rights by prior appropriation. The doctrine holds that the first person to use water from a stream earns the right to that amount of water in perpetuity: "First in time, first in right." A downstream user who got to the stream first could force later-arriving upstream users to give up water that flowed by their door. By 1922, prior appropriation had largely replaced riparian rights across the West. And no state had been using Colorado River water longer than California.

The Court ruled that when two states have a common water source and recognize prior appropriation, which all seven Colorado River basin states did, they are bound by the doctrine. In the early 1920s, outside of a few fledgling cities like Denver and Salt Lake, the upper basin states were sparsely populated and far from being able to take their full allocation of Colorado River water. By the time they were able to do so, California might already be using the water and, given the Court's endorsement of prior appropriation, could keep it.

In 1923, Congressman Phil Swing, former chief counsel and lobbyist for the Imperial Irrigation District, introduced a bill to authorize the high dam on the lower river and the All-American canal that his constituents had long sought. Senator Hiram Johnson of California later co-sponsored. But if Congress approved the bill before the states had agreed on how to divide the river, the Court decision on prior appropriation would surely lead to a lengthy, expensive fight. Congress had paved the way to a negotiated agreement the year before by approving a Colorado River Commission and appointing Secretary of Commerce Herbert Hoover to chair it. No one could be confident about the commission's chances of success—nothing stirs the blood like a dispute over water. The very word "rival" comes from the Latin *rivalis:* "one living on the opposite bank of a stream from another."[6] After months with nothing to show for their effort, the commissioners agreed to gather in November 1922 at Bishop's Lodge, near Santa Fe, a congenial setting for a compromise.

The stubborn LaRue decided to seize the opportunity. He invited members of the commission, along with engineers from the Geological Survey and the Bureau of Reclamation, representatives of the Southern California Edison Company, the Mormon Church, and the State of Utah, to see the canyons of the Colorado River for themselves. The party traveled overland through beautiful Capitol Reef in central Utah. From there they made their way to a spot 110 miles above Lee's Ferry, where Hall's Creek enters the Colorado in Glen Canyon. LaRue and his boatmen motored upstream from Lee's Ferry, picked up the party, and floated them downstream. All the while, LaRue regaled his captive audience with the advantages of Glen Canyon as the site of the first large dam on the Colorado. The party disembarked at Lee's Ferry and traveled overland via the Hoover Dam site to Flagstaff, where they boarded the train for Santa Fe.[7]

The voyagers could not help noticing that in the entire stretch they rafted, Lee's Ferry was the only safe place to leave the river. A sailor who missed that lone opportunity would have to ride the river three hundred miles to the mouth of the Grand Canyon—if he survived the scores of rapids along the way. Everyone knew that three of the crew from Powell's first voyage had tried to find another way out of the Grand Canyon, only to perish on the plateau. Lee's Ferry was the natural dividing point between an inaccessible upper river and an inaccessible lower one.

The boaters basked in the same peaceful, rapid-free Glen Canyon that had enchanted Powell and his crew. They visited Anasazi ruins and hiked the six miles up to Nonnezoshi, as the Navajo call their sacred Rainbow Bridge. In his autobiography, one of the voyagers, John Widtsoe, remembered the awe-inspiring trip "down the river between picturesque rock walls sometimes more than two thousand feet high."

Sleeping on the sweet sand was full of romance. Darkness closed in with multiplied whisperings of silence. Shadows climbed up the vertical side of the narrow canyon in huge and awesome forms. Then with the first light of morning, the canyon rioted with changing color. As the sun rose, though the canyon remained in the shadows until the sun was high, a glorious moving, colorful, ever-changing panorama played on the canyon walls. We watched it in breathless wonder.[8]

Figure 7. Colorado River Commission, 1922, with Delph Carpenter (*left rear*) and Herbert Hoover (*second from right, front row*) (Colorado State University, Water Resources Archive, Archives and Special Collections)

One of the commissioners was Delph Carpenter, an accomplished Colorado lawyer and western water expert. Carpenter was a great compromiser who believed that any group in possession of the facts and honestly bent on reaching an agreement could settle any river question.[9] History casts doubt on this thesis, but at Santa Fe, Carpenter made it come true. Several years later, President Hoover wrote to Carpenter, by then severely afflicted with Parkinson's disease, to say, "That compact was your conception and your creation, and it was due to your tenacity and intelligence that it succeeded."[10]

Carpenter came up with a way to break the stalemate. He proposed that instead of apportioning water state by state, which would surely invite invidious comparisons, the river should be divided into two multi-state basins, with each getting an equal share of water. How much water each state within a basin should receive would be decided later. Carpenter's proposal thus solved one problem only to create another, but it did allow the commissioners to show that they had accomplished something. They divided the river at the only feasible spot: Lee's Ferry.

What appeared to be the commissioners' easiest task would turn out to have the longest-lasting consequences. To divide the flow of the river into equal halves, they had to know what number to divide. For two reasons, it was easy to get the flow of the river wrong and they did. First, since there was no river gauge at Lee's Ferry, the flow there had to be interpolated from gauges up and down the river—and there were few of them. Second, records had been kept only since 1896, providing a short, twenty-six-year sample of a river that, although no one knew it at the time, had been running for at least five million years. And indeed the sample turned out to give a seriously wrong result. The years that preceded the commission, we now know, were wetter than the long-term average. From 1906 to 1921, the flow averaged 18.5 MAF and none of those years had less than 12 MAF. A scrutiny of commission minutes by R. Eric Kuhn leads him to conclude that the commissioners believed the river at Lee's Ferry carries an average of at least 17.2 MAF each year.[11] In allocating each basin 7.5 MAF, the commissioners must have felt that they had provided a comfortable cushion, but actual flows during the rest of the twentieth century were to prove them wrong.

Recognizing that the flow of the river varies markedly from year to year, instead of allocating an annual amount to each basin, the Colorado River Compact wisely defined a ten-year rolling average: "The states of the upper basin will not cause the river at Lee Ferry to be depleted below an aggregate of 75,000,000 acre feet for any period of ten consecutive years reckoned in continuing progressive series."[12] As we will see, between requiring the upper basin to *deliver* 75 MAF every ten years and requiring it "not to cause the river . . . to be depleted" lies a subtle but vital distinction. The compact did not define "cause" or "deplete," leaving the task to some twenty-first-century court.

Reflecting the underdeveloped status of most of the Southwest, the compact favors agriculture over cities and irrigation over power generation. Any future obligation to Mexico, which previously had gotten 100 percent of the lower river, would come from the surplus. If there were no surplus, the two basins would share equally in the delivery of water south of the border. Even though for its last hundred miles or so the river flows through Mexico, Hoover did not believe the Mexicans had any rights to Colorado River water. He accepted the language only in the interest of "international comity." But Mexico still fared better than Native Americans, whose rights the future president of all the people said were "negligible." In what Hoover called the "wild Indian article," the compact said merely, "nothing . . . shall be construed as affecting the obligations of the United States of America to Indian tribes." How much water is owed Indian tribes will also have to be adjudicated in the coming decades.[13]

Congress, which had been shocked to hear that LaRue believed there was too little water in the Colorado River to irrigate all the land in its basin, called the hydrographer to testify. LaRue might have used diplomacy and tact to persuade the congressmen of the need for a unified plan for developing the Colorado River, including a linchpin dam at Glen Canyon. Instead, he responded acerbically to questions and showed more concern with building a record for "any person who may read it" than in persuasion. Engineer Hoover followed LaRue, testifying that "logic drives us as near to the power market as possible, and that therefore takes us down to into the lower canyon."[14] When questioning revealed that no test drilling had been done at Glen Canyon, LaRue had lost the day. But the mulish river man continued to agitate and lobby for his plan. Finally a fed-up director of the Geological Survey ordered him to observe silence on the matter of Boulder Dam. Rather than wear the gag, the proud LaRue resigned.[15]

California did not mind that Carpenter's Solomon-like solution failed to specify how much water the states within each basin were to receive. As long as there was surplus water in the river, the Golden State could take as much as it wanted. California's representatives knew that when the time came, they had the political power to ensure their state at least its fair share. Nevada's needs were minute, but Arizona, long fearful of its giant neighbor, was understandably wary. The key question was how the water in Arizona's own rivers, especially the Gila, would count. Arizona's representative insisted that the state should receive its fair share of the lower basin allocation *plus* the water from its own rivers, which amounted to over one million acre-feet. When the other states responded that the Gila should be subtracted from Arizona's allotment, the state refused to sign the compact.

Congress had previously decreed that for the compact to become law, each of the seven basin states had to ratify. If Arizona continued to balk, the states would quickly begin to fend for themselves. Then the commission's work would have been in vain, and Congress would likely have to step in and dictate to the states how much water each should receive. Carpenter then called Arizona's bluff: he proposed that to ratify the compact, only six states need sign. California agreed, but on condition that the federal legislation include a high dam on the lower river. The representatives of the upper basin also agreed, but countered with their own condition: the legislation should specify the maximum amount of water that California would be allowed to take from the river.

With the compact in place, in 1928 Congress passed the Swing-Johnson bill, now known as the Boulder Canyon Project Act. California could take 4.4 MAF plus half of any surplus. Arizona would get 2.8 MAF plus the other half of the surplus, plus the water from the Gila. Nevada would receive 300,000 acre-feet, more than enough for the nearly vacant state. Since the act specified how much water each state in the lower basin could take, it nullified prior appropriation in the Colorado River basin.[16] The act placed the responsibility for the entire project in the hands of the secretary of the Interior. President Hoover signed the legislation on 25 June 1929. In three years, he would be out of a job and the nation would have fallen into desperate straits. By then, not only the Bureau of Reclamation but the entire nation would need a grand project—the bigger the better—to provide jobs, lift spirits, and show that even a great depression could not bring down the United States of America. By 1930, many factors had coalesced to give federal reclamation a second chance. The locus of the bureau's rejuvenation would lie just below the mouth of the Grand Canyon, where Arthur Powell Davis had long dreamed of building a great dam.

SEVEN

Lonely Lands Made Fruitful

AT 726 FEET, HOOVER DAM would be twice as high as the record holder at the time, the bureau's Arrowrock Dam in Idaho. The monolith would contain 4.4 million cubic yards of concrete, one-quarter more than the volume of the Great Pyramid at Giza. It would weigh nearly seven million tons and generate its own earthquakes. Hoover Dam proved such a marvel that thirty years later, engineers used the same design and techniques to build Glen Canyon Dam. Indeed, Hoover Dam was the first of tens of thousands of large dams across the world, concrete evidence that the fastest route to first-world status is a large hydropower dam.

The first obligation in building a dam is to make it safe. On 12 March 1928, as the bureau was drawing up its plans for Hoover Dam, forty miles north of the city, Los Angeles's St. Francis Dam collapsed. The disaster cost more than six hundred lives—the exact number was never known—and left bodies strewn along the coast from Ventura to San Diego—forty schoolchildren among them. Hoover Dam would be far larger and, should it fail, far more destructive. On 29 May, Congress ordered the secretary of the Interior to oversee a study of Boulder and nearby Black canyons, both on the lower Colorado River beyond the mouth of the Grand Canyon, and recommend one for construction. When the report came in six months later, Black Canyon was the choice. The site had a more solid and therefore safer bedrock foundation, was easier to get to, required less sediment removal, could impound more water and therefore offered more power generation and flood protection. President Hoover's secretary of the Interior, Ray Lyman Wilbur, arrived in Las Vegas on 17 September 1930 to drive a large silver spike to inaugurate the big project. To the amusement of the many ex-miners and railroad men in attendance who had actually driven spikes, in his first few at-bats the high-collared bureaucrat swung and missed. After finally connecting with the spike, Wilbur surprised his audience by announcing that the dam would not go up in Boulder Canyon, where everyone had expected it would, but in Black Canyon. Moreover, instead of being called Boulder Dam, "in accordance with many requests," it would be named Hoover Dam. As soon as he took office three years later, Franklin Roosevelt's irascible secretary of the Interior, Harold Ickes, changed the name to Boulder Dam. But in 1947 Congress changed it back to Hoover Dam. President Harry S. Truman, perhaps thinking of the virtues of preserving presidential legacies, signed the bill and Hoover Dam it has been ever since.

The Bureau of Reclamation wanted to give the job of building the huge dam to a private construction company, but none were large enough. Several local western firms had built roads, laid railroad lines, put up modest buildings, and the like. None could build a dam the size of Hoover by themselves—or raise the $5 million bond that the government required. Several in combination might be up to the job, but the kind of man who had built a small western construction company from scratch and stayed in business even as a national depression began was not by nature inclined to submerge his identity. Yet, in the end the lure of the biggest project in the nation's history overcame the independent streak of this can-do breed.

The idea of forming a conglomerate came from Harry Morrison of the Morrison-Knudsen construction company out of Boise, named for the two engineers who had quit the bureau to go into business for themselves. They had already partnered with the Utah Construction Company, who agreed to join. Charlie Shea had built tunnels and sewers in Los Angeles and saw no problem with raising the bond. He liked to say, "I wouldn't go near a bank unless I owed them at least half a million dollars—that way you get respect."[1] Warren Bechtel, a successful San Francisco contractor, was amenable and persuaded a road builder from Oakland, Henry Kaiser, to come in with him. In all, eight companies agreed to band together to take on the big job.

In February 1931, representatives of the eight met at the Engineer's Club in San Francisco. Borrowing their name from the Tongs then running the bay city's Chinatown, and defying arithmetic, the group called itself Six Companies. The sine qua non of its effort was to submit the low bid and win the contract. Bureau of Reclamation engineers had estimated the cost of the project to the last penny and said they would accept no padding. To prepare its bid, Six Companies needed a man who had built dams and knew how the bureau operated. Harry Morrison had the perfect candidate: a man who had already designed a high dam on the lower Colorado.

As a student at the University of Maine in 1904, Frank Crowe had heard an engineer from the new Reclamation Service describe its vision of reclaiming the arid west. The student, whose nickname would become "Hurry-Up Crowe," wasted no time in asking the speaker for a job. He got it and quickly rose to general superintendent of construction in the service. As early as 1919, Crowe and Arthur Powell Davis were working up a cost estimate for a high dam on the lower river. By 1924, Crowe was assisting in the dam's design.[2] Then in 1925, the bureau announced that it would contract out its dam construction. Crowe, who wanted to build dams himself, resigned and went to work for Morrison-Knudsen. "I was wild to build this dam," Crowe would later recall. "I had spent my life in the river bottoms, and [Hoover] meant a wonderful climax—the biggest dam ever built by anyone, anywhere."[3] Under Crowe's direction, Six Companies' bid of $48,890,995 came within $24,000, or 0.05 percent, of the bureau's estimate.[4] The firm won the competition and appointed Frank Crowe superintendent of construction.

Six Companies would go on to build Bonneville and Grand Coulee dams on the Columbia, the foundations for the Golden Gate and Bay bridges, and many large projects at home and abroad. Eventually its component firms would go their own way, as Bechtel, Kaiser, Morrison-Knudsen, and the like. Thus Hoover Dam not only launched the era of mega-dams, it spawned the mega-construction company.

Building an enormous concrete dam at the bottom of a narrow canyon in a torrid wilderness brought special engineering and construction challenges. To withstand the pressure of the giant lake—over nine trillion gallons of water—the dam would use the principle of the arch. To anchor the dam, its base would be several times thicker than its crest.

Concrete can be formed into whatever shape designers wish, providing structures for the power plant, waterworks, galleries, access tunnels, and the like. But it cannot be poured underwater. To keep the dam site dry, the bureau would build a cofferdam upstream from the site to divert the river into tunnels drilled through the canyon walls. Reconnaissance geology had located plenty of rubble near the dam site that could be used to build the cofferdam, but would the trucks be able to dump it fast enough to get ahead of the wild river? Twenty-five years before, the Southern Pacific Railroad had spent two years dumping carload after carload of gravel into the lower Colorado before it finally succeeded in turning the river back into its original channel. Hoover Dam was on a fast track, with large bonuses awaiting an on-time and on-budget delivery. Six Companies did not have two years to spend dumping gravel before it could even begin.

Assuming that the cofferdam succeeded in diverting the river into the tunnels, how could the construction company pour millions of tons of concrete? Not all at once—to produce and pour that much in one fell swoop was mechanically impossible. Even if it could be done, as the concrete hardened, chemical reactions would generate so much heat that the mass would take 125 years to cool.[5] Somehow the designers had to find a way to build the dam out of small batches of concrete that could be mixed on the canyon rim, sped down to the dam site on the floor, and poured into forms. Each block had to cool rapidly enough to allow the next batch to be poured on top of it. From hundreds of these smaller blocks, the great dam would rise. But calculations showed that even the smallest feasible batches of concrete would still take too long to cool.

The bureau gave the construction company thirty months from the time work started in May 1931 to complete the diversion tunnels. At first, without roads or electrical power in Black Canyon, men and equipment had to float down the river and moor at the excavation site. But soon the workers had built roads and suspended a cableway from a tower on the Nevada side down to the canyon floor on the Arizona side. The cableway was key. Without it, not enough material and equipment could be brought to the dam site fast enough. Electrical power arrived through a line from San Bernardino 220 miles to the west, the same line that would carry Hoover Dam hydropower back to Southern California.

At first, the tunneling crews drilled one hole at a time into the rock wall, filled it with dynamite, and shot it off. The work speeded up when the drilling jumbo arrived. Built on a ten-ton truck, this Rube Goldberg contraption had three platforms, one above the other, from which a crew of fifty men drilled twenty-five to thirty holes at a time. Once enough had been drilled on both sides of the rock face, the crew filled the holes with dynamite and shot it off, advancing the tunnel by fifteen feet. Then carbon monoxide-spewing trucks hauled sixteen cubic yards of the rubble at a time downstream and piled it up for later use. The diversion tunnels, two on each side of the canyon, began a thousand feet upstream of the dam site. Curving through the dark, adamantine igneous rock of Black Canyon, they emerged three-quarters of a mile below the dam. At fifty-six feet in diameter, the tunnels could hold six New York City subway cars parked side by side. Eventually the workers removed more than 1.5 million cubic yards of solid rock from the canyon walls.[6] Finally, they installed a three-foot concrete lining to protect the tunnels from erosion.

On 13 November 1932, enormous explosions blasted open the upstream ends of the two tunnels on the Arizona side. The giant maws were ready to take the river—if it could be forced from the path it had followed for millions of years. A line of waiting trucks pulled out on a trestle bridge just below the tunnel entrances and began to dump load after load of rock and gravel. The river rose against the barrier, roiled and swirled, and then, finding the easier path that the dam-builders had prepared, began to flow off into the tunnels. To keep the diverted flood of water from washing back upstream, crews built a second cofferdam below the dam site.

Thousands of years of weathering had left the rock face of Black Canyon cracked and broken. Before concrete would safely adhere, men had to get to the canyon wall, remove the dangerous pieces, and smooth the surface. Since they could not climb up the rock face from below, the workers had to descend on ropes. Suspended in mid-air, the high scalers had the most dangerous, the most exhausting, and therefore the most prestigious job on the dam. They came from many backgrounds—sailors, Native Americans, circus acrobats. What they had in common was a willingness to hang suspended hundreds of feet up, on a slender rope, drill a hole, set off a dynamite charge, and swing out of the way—Tarzan-like—before the charge went off.

Once the high scalers had descended to the work level, men on the rim lowered them a forty-four-pound jackhammer. Negotiating a maze of air hoses, current-bearing electrical cables, and bundles of drill steel, the high scalers made their way to their assigned spot on the wall and began to drill. Rock and tools fell on them, so they made themselves hard hats out of cloth soaked with coal tar. Surprisingly, the makeshift chapeaus worked—several men were struck on the head hard enough to break their jaws, yet the blow did not fracture their skulls. Eventually Six Companies saw fit to provide proper hard hats.[7]

To be a high scaler, a man had to be a daredevil if not downright crazy. When the foreman turned his back, the Hoover Dam aerialists would push off hard to see who could sail farthest from the canyon wall or perform the most acrobatic stunt. At one spot, a boulder protruded from the Arizona side of the canyon. To get around the obstruction, a workman would wrap his legs around the waist of the high scaler, who would then push out and swing around to the other side of the boulder, where he would deposit his passenger and return for the next. This went on twice a day for several weeks. A bureau engineer named Burl Rutledge once lost his balance and fell from high on the canyon rim in what would surely be a fatal drop. Twenty-five feet down, high scaler Oliver Cowan, who had heard the noise of Rutledge's fall, pushed out and grabbed him by the leg. Seconds later, another scaler swung over to lend a hand. The two held on until the lucky engineer could be pulled to safety.[8]

With the cofferdams blocking the river and the canyon walls nearly ready to receive concrete, it was time to prepare the foundation. To anchor the dam in solid bedrock, crews had to remove the silt that had accumulated in the bed of the river. Forty feet down, at the bottom of the sediment pile, they came upon a milled piece of timber.[9] Since Indians did not saw lumber, the plank could only have floated downstream after the white man arrived in the drainage basin of the Colorado River. That meant that within the preceding seventy-five years or so, the river had run so high that it had scoured its bed clean. When the high water subsided, the river deposited new sediment and the wooden plank. The lonely board was a reminder of how much high water Hoover Dam might need to hold back and how much silt the river carried.

Crews now confronted the problem of how to pour concrete in batches and cool it fast enough so that the huge dam would rise as a single, unified mass. At Owyhee Dam in Idaho in 1931, engineers had confronted the same problem.[10] There they hit on the ingenious solution of running cold water through coils of hollow pipe embedded in the concrete. The men used the same device at Hoover Dam, running first river water through the coils and then ice water produced at a refrigeration plant on the rim. The chilled water quickly cooled the blocks; workers then cut off the pipes and sealed them with grout.[11]

On 6 June 1933, nearly eighteen months ahead of schedule, the first bucket of concrete, mixed at a plant on the rim, sped down the cableway to a waiting wooden frame. It was a tricky business. If the wet concrete took too long to arrive at the bottom of the canyon, it would begin to set up in the dump bucket and have to be chipped out by hand, delaying the next batch. Once a block had been poured, it had to be kept wet so that it would bond with the next pour. Down in the hole, a crew of twenty men continuously sprayed water to keep the surface of the newly poured blocks moist. After a block cured, workers shoved grout into the openings; as the blocks expanded they compressed the joints into a single continuous face.

The original design of Hoover Dam had not included power generation. When the bureau decided to add a power plant, the question was where to put it. Locating the plant on the side of the canyon would have required more expensive and time-consuming excavation. Therefore the bureau elected to build the plant at the downstream toe of the dam. That in turn meant that the spillways could not run over the dam crest, for then high water would descend and damage or destroy the powerhouse. Instead the designers placed the spillways high on the canyon wall above the dam. From there the spillway chutes plunged down to intersect the no-longer-needed diversion tunnels. The Hoover spillways were engineered to pass 400,000 cfs. The same technique would be used at Glen Canyon, though the bureau designed the spillways there to pass 276,000 cfs.

Hoover Dam qualified not only as a great engineering feat but also as a monumental work of art. Bureau engineers, not given to flights of artistic fancy, had planned to embellish the concrete with the usual eagles, cornices, and the like. Instead, bureau officials hired architect Gordon B. Kaufman and artist Allen True to design nearly every external feature of the dam and its appurtenances. Rather than hanging ornaments here and there, Kaufman converted the structures themselves into classic examples of modernism and Art Deco design. True covered the floors of the interior with unforgettable terrazzo patterns using motifs and colors—black, white, green, and ocher—that were adaptations of Pima, Acoma, and Mimbres designs. Naturalized American citizen Oskar Hansen added two giant, eagle-like sculptures, the *Winged Figures of the Republic*.[12]

On 29 February in the leap year of 1936—two years ahead of schedule—the construction company handed Hoover Dam over to the Bureau of Reclamation. Remarkably, the biggest construction job ever undertaken had been completed in only five years, in a country in the midst of an awful depression, in a region of almost unbearable working conditions. The achievement was so significant that Franklin Delano Roosevelt, whose polio affliction made travel difficult, journeyed all the way to the remote site to dedicate the dam. FDR told his listeners that Hoover Dam was "an engineering victory of the first order—another great achievement of American resourcefulness, skill and determination."[13]

But the victory had come at a terrible cost.[14] The official Bureau of Reclamation death count numbered 114 men. But that toll does not include the women and children who died in the squalor of Ragtown, the decrepit shantytown where workers were forced to live before government housing became available in Boulder City. The two most poignant deaths may have been those of J. G. Tierney, who drowned on 20 December 1922 while surveying the future dam site, and, thirteen years later, of his son Patrick, who fell from one of the intake towers and drowned. During that deadly summer of 1931, the air in the diversion tunnels was pure poison. One man remembered, "We could tell by looking at the lights in the tunnel. If they had a big blue ring around them, we would know the gas was getting pretty rough in there." On one bad night, of the seventeen men who went into the tunnels, thirteen wound up in the infirmary. "They were hauling men out of those tunnels like cordwood," said one observer.[15]

Those who worked outside did not have to contend with poisoned air—only air that was unbearably hot—and heat can prove as deadly as carbon monoxide. In July 1931, the high temperature averaged 119 degrees and the low 95 degrees—but that was on the rim. Down in the airless box of the canyon, surrounded by black, heat-absorbing rock, amidst blowing dust and beneath vertical walls that offered no shade, trying to hang on to a bucking jackhammer, a man faced temperatures that could approach 130 degrees. In one thirty-day period that summer, fourteen men died of heat prostration and two drowned. Nowhere was safe: on 26 July, three women died in the dreadful Ragtown camp.

On 7 August 1931, the workers went out on strike, demanding clean water and flush toilets at their houses, ice water at the work site, and adherence to the mining laws of Arizona and Nevada. Knowing that for every disgruntled dam worker, another man in Ragtown or Las Vegas would be glad to take his place, Six Companies rejected each demand and called in strikebreakers with clubs and guns. In less than a week the strike was broken. Within one year, twenty-seven men had died on the site. They fell; they drowned in water tanks and in the river; falling rocks, trucks, and power shovels crushed them; power lines electrocuted them; cables lashed them to death; premature explosions blasted them to pieces. On the big project went, ahead of schedule. At another iconic and dangerous giant American project, the Empire State Building, completed the year construction of Hoover Dam began, a total of five people died.[16] During the years it took to construct Glen Canyon Dam, by one account, seventeen workers lost their lives.[17]

For bringing the big project in early, Frank Crowe received a bonus that dwarfed his $18,000 annual salary. He got $350,000, equivalent to about

$5,000,000 today.[18] He went on to build two smaller dams downstream on the Colorado, and then Shasta on the Sacramento River in Northern California.[19]

The social engineering at Hoover Dam, or the lack of it, had proved deadly. Determined not to repeat its mistakes, to house the workers at Glen Canyon thirty years later, the bureau built an entirely new town and had it ready when the work began. Page, Arizona, became a model American small town, one sustained to this day by the appeal of Lake Powell as a recreation site. As part of the plan for the town, the bureau set aside land for all major denominations. Church Row remains the centerpiece of Page.

Whatever the fate of their reservoirs, Hoover and Glen Canyon dams will stand for millennia, tributes to the brave men and their families who risked their lives to build these giant emblems of the twentieth century.

Many arrive at Hoover Dam and Lake Mead after a visit to Lake Powell and the Grand Canyon. Motoring west across the high Arizona desert, they swing around a curve to find, crouching amidst brown and black rock, a huge, gleaming white dam and a sapphire lake. The first impression of startling contrast between dam, rock, and water gives way to amazement and appreciation. Even from today's prosperity, one can understand how Hoover Dam inspired a depressed America. As one of the original plaques reads, "It is fitting that the flag of our country should fly here in honor of those men who, inspired by a vision of lonely lands made fruitful, conceived this great work and of those others whose genius and labor made that vision a reality."[20]

The gleaming dam, the sapphire lake, the inspiring inscriptions, plaques, sculptures, and exhibits—all give the visitor a sense of a great mission accomplished, of desert made fecund, of a mighty and benevolent government shepherding its people. The visitor has no way to learn how well the benefits of Hoover Dam subscribe to the vision of Thomas Jefferson, who said that "the small land-holders are the most precious part of the state."[21] Or to the concern of John Wesley Powell for the "poor man." Or to the intent of Francis Newlands that reclamation law "guard against land monopoly and . . . hold this land in small tracts for the people of the entire country . . . [and] to give to each man only the amount of land that will be necessary for the support of a family."[22]

Figure 8. "Boulder Dam, 1942" by Ansel Adams. Later the name reverted to Hoover Dam. (National Archives and Records Administration, Records of the National Park Service)

The Boulder Canyon act gave the farmers of the Imperial Valley the high dam they sought, with its attendant flood control and 28 million acre-feet of stored water. All the act required of them was to pay for the cost of the All-American Canal, in effect a giant ditch. One would have thought that the irrigators, having benefited so greatly at taxpayer expense, would then have been willing to obey federal reclamation law.[23] After all, the Boulder Canyon Project Act said that valley residents were to pay for the canal "in the manner provided in the reclamation law."[24] Surely that implied that no Imperial Valley landowner could hold more than 160 acres, 320 for a couple, and that the owner would have to reside on the land. Those with parcels larger than the acreage limit would have to sell off their excess and, if they lived elsewhere, move onto their property.

Even as the delegates negotiated the Colorado River Compact, farms in the Imperial Valley were growing beyond the acreage limit. According to western water historian Norris Hundley, Jr., in the 1920s Imperial Valley farms often comprised 300 to 700 acres, and some reached 3,000 acres.[25] By 1974, seventy-three valley farms comprised over 1,000 acres and about the same number exceeded 2,000 acres.[26] The battle over the acreage limitation lasted for fifty years and during all that time, like a prisoner on death row living on through appeals and legal maneuvers, Imperial Valley farmers never had to meet any limit.

The ability of the landowners to ignore the acreage maximum for so long rested on the slender reed of a letter from Hoover's secretary of the Interior and spike-driver, Ray Lyman Wilbur. His assistant, Northcutt Ely, had at first concluded that there was "nothing to do but enforce [the acreage limitation] unless the Imperial Irrigation District can get new legislation."[27] The chief counsel of the Bureau of Reclamation confirmed the opinion. But the Imperial Irrigation District mounted an intense lobbying campaign to prevent the valley's water contract from mentioning any acreage limit. If one were included, they warned, valley voters might not ratify the contract. As the time neared for the Hoover administration to leave town, Ely reversed himself and drafted a letter for Secretary Wilbur that exempted the Imperial Valley from the acreage limit. On 24 February 1933, eight days before FDR's inauguration, Wilbur signed the letter. His stated grounds were that the limit did not apply because the valley would not use new water, only the same water it had been using before the contract. According to Wilbur, that gave valley irrigators a vested right to the water.[28] The Wilbur letter was at best an informal opinion without force of law. Nevertheless, it provided a quasi-legal cover that allowed the district to litigate endlessly while it continued to ignore the acreage limit. After five decades of legal battles, in June 1980, the Supreme Court overruled an appeals court decision and found for the large landowners in the Imperial Valley. Serving as counsel for the district was Northcutt Ely.[29]

The second requirement of the original reclamation law was that landowners reside on and farm their own land. Like the acreage limit, the restriction has been honored in the breach. Today, some Imperial Valley farms resemble southern plantations more than small homesteads. Most of the seasonal workers live across the border in Mexicali, from where they are picked up and returned after working for a day at minimum wage. For those who live in the Imperial Valley, the American dream is not even within sight. Of California counties, Imperial has the highest resident unemployment, the highest percentage of Hispanic residents, the highest percentage of people on welfare, and the greatest number of families below the federal poverty line. The county also has one of the highest rates of childhood asthma in the nation.[30]

In 2000, the Imperial Valley had 575,000 acres in cultivation, one-third of it in alfalfa, making the valley the number one alfalfa-growing region in the world.[31] Most of the fodder is baled and shipped to California dairies, but some is compressed and sent overseas. Alfalfa is an especially thirsty crop that accounts for nearly 20 percent of California's agricultural water use, though it contributes only 4 percent of the state's farm revenue.[32] Imperial Valley farmers recently paid less than $15 per acre-foot, one-twentieth the price of water in nearby San Diego.[33]

The abandonment of the acreage limit and the residency requirement allowed absentee agribusinesses to amass spreads of tens of thousands of acres, directly repudiating the intent of the reclamation act. Water districts like Imperial and Westlands in California's San Joaquin Valley have grown more powerful than the counties that contain them. The railroad barons have gone on to the end of the line, replaced by the barons of water. Before long, water will be more valuable than the crops it irrigates, prompting water districts to sell their rights to cities and keep the profits.

But back in 1936, this future lay decades ahead. The bureau's immediate concern was to parlay the success of Hoover Dam into more dams, a larger budget, and bureaucratic longevity. As E. C. LaRue had insisted, the Colorado River upstream from Hoover Dam held many excellent sites to help the agency achieve all three goals.

PART THREE

River of Controversy

EIGHT

Natural Menace Becomes National Resource

AS THE GREAT DEPRESSION extended its tentacles across America in the winter of 1930–31, farmers had a hard time repaying their loans. The reclamation fund, supposed to have reached steady state years before, had begun to dry up, threatening the bureau's ability to finish the projects it had under way. Just when the agency was about to lay off 1,500 workers, Congress intervened with a $5 million loan. But the legislators simultaneously declared a moratorium on repayment, which could only necessitate future bailouts. Sure enough, in 1933 the government had to lend the fund another $5 million, one more step in a decades-long shift of financial responsibility for federal reclamation from farmers to the national taxpayer.[1] Then, just as its finances had sunk to a new low, the Bureau of Reclamation experienced one of the more dramatic reversals of fortune of any government agency outside of wartime. Hoover Dam and the New Deal would make the Bureau of Reclamation one of the most, if not the most, powerful agencies in American government.

After the election of Franklin Delano Roosevelt, federal reclamation aspired to more than reclaiming land. Now the bureau was a key agency in a program intended to power the nation and put it to work. Before big dams generated electrical power, they would generate jobs. The bureau's projects became such a priority that in 1934 the Public Works Administration awarded it $103 million for construction, more than ten times the agency's average budget over the preceding thirty years.[2] But since new dams and irrigation projects take time to return revenue, the reclamation fund continued to pay out more than it recouped. In 1937, the fund received only $1.2 million, one-ninth the amount it had gained in 1908, when federal reclamation was just getting under way. The bureau naturally feared that once the Depression ended, so would its large construction budgets, forcing it to fall back on the rapidly dwindling fund. In 1938, Congress stepped in again, this time with money from something called the Naval Petroleum Fund. The following year, in an effort to put the bureau's finances on a sounder long-term footing, Congress passed the Reclamation Project Act of 1939. Among other things, the legislation extended the loan repayment period to fifty years, a tacit admission that many reclamation projects would never pay for themselves. In 1939, fifty years was well above the remaining life expectancy of a thirty-year-old American white male.[3]

The arrival of World War II ended the Great Depression, set the nation on a war footing, and elevated hydropower to a matter of national survival. In 1941, the bureau completed Grand Coulee Dam on the Columbia River, the world's largest hydropower dam and, in the words of folksinger Woody Guthrie, the "mightiest thing ever built by a man." Meanwhile, over in eastern Colorado, people felt that federal reclamation had passed them by. The water they coveted lay on the other side of the mountains, on the west slope of the Front Range, providing bureau engineers with just the kind of challenge they enjoyed. The agency designed a package of dams, reservoirs, and pumps that would convey water from the western slope of the Rockies via a thirteen-mile tunnel drilled through the mountains. Water exiting the tunnel on the east side would flow down the Big Thompson River and from there on to fields and towns. The Colorado–Big Thompson, as the project became known, demonstrated just how ingenious the bureau had become in overcoming natural obstacles and taking water to wherever it wanted.

But the upper basin states had a greater concern than how to get water across the Front Range. The source of their worry lay a thousand miles downstream, in a state whose unslakable need for water had given it what historian Norris Hundley, Jr., has called "The Great Thirst."[4]

When the commissioners signed the Colorado River Compact in 1922, California had three and a half million people. At the time, the population of the four upper basin states (New Mexico, Colorado, Utah, and Wyoming) totaled less than two million. By 1930, California had *added* two million people. As long as the upper basin states remained relatively undeveloped, the surplus in the lower river was significant and California would surely get used to using all of it. The upper basin states feared that to avoid giving up the surplus water, the powerful California bloc in Congress might try to revisit the compact—but the delegation would not even have to go that far. Having the votes to deny authorization of upper basin water projects, California legislators could kill them without the brouhaha that an attempt to revise the Law of the River would produce.

The Dust Bowl years reminded everyone that a severe drought could threaten the ability of the upper basin to deliver its rolling, ten-year obligation of 75 MAF. In that event, California would surely issue a "compact call," requiring the upper basin to supply the missing water by reducing its consumption or drawing down its reservoirs, or both. To prevent that possibility—to gain the kind of drought insurance that Lake Mead provided the lower basin—the upper basin needed its own large reservoir. The upriver states would not need to draw even a drop of water from the reservoir—its only purpose would be to guarantee that the upper basin could meet its downstream delivery debt under the compact. Therefore the bureau could build the dam at the very foot of the upper basin, even at LaRue's old Glen Canyon site, only a few miles above Lee's Ferry.

If the upper basin did not yet have many people, it did have an abundance of canyons. From the Gate of Lodore on the Utah-Wyoming border to the mouth of the Grand Canyon, the Green and Colorado rivers have carved one chasm after another. North to south, they include Lodore, Desolation, Gray, Cataract, Glen, Marble, and Grand. Most contained at least one good dam site and the stretch of river between Marble and Grand canyons held several. The San Juan and other mainstem tributaries added their own sites to the list. But Nature had limited how much benefit dams in the upper basin could provide.

The farther upstream one travels, the higher the elevation, the colder the climate, and the shorter the growing season. The Big Horn basin of Wyoming, for example, which produces a sizable fraction of the state's crops, stands at an average elevation of 4,500 feet. The Uinta basin in Utah and the western slope of the Colorado Rockies both average a mile or more. At these altitudes, with their brief growing seasons, the most profitable crops tend to be alfalfa, spring wheat, barley, oats, corn, and pasture grasses. These thirsty plants require large-scale irrigation but return relatively less money per acre than, say, fruits and vegetables. By the time the bureau was ready to propose dams in the upper basin, many of the crops that would grow on the reclaimed land were already in surplus. It did not take an agronomist to see that irrigation projects on these high plateau lands were bound to lose money. Even in more favorable climes, farmers on bureau projects had been unable to repay their interest-free loans, leading Congress to extend the repayment period repeatedly. A farmer growing alfalfa where it can snow in early June and again in late August would be bound to do worse. Since the bureau could not change the altitude and climate of the upper basin, it changed its accounting method.

—

In 1946, the bureau published the report that eventually led to a high dam at the foot of Glen Canyon, called *The Colorado River: A Natural Menace Becomes a National Resource.*[5] Nicknamed the "Blue Book" for its cover, the report described 134 separate projects in the upper basin, almost as many as the four states have counties. If the bureau could implement enough of the projects, it would bring the Colorado River completely under man's domination, turning it from a wild thing into an automaton.

But whom did the Colorado River menace? Not the people on the lower river, below the Grand Canyon: Hoover Dam protected them. For a thousand miles upriver, there was hardly anyone to menace. The first two settlements of any size were Moab, Utah, on the Colorado and Green River, Utah, on the stream of the same name. The bureau had not previously shown much concern for the two little river towns. LaRue would have submerged both under a giant lake impounded at the confluence of the Green and the Colorado. To residents of Moab and Green River, the real menace might have appeared to be not the rivers, on whose banks they had purposely settled, but the Bureau of Reclamation. For hundreds of miles above the two towns, the rivers flowed in isolated canyons. To call the Colorado River a menace was mere propaganda, but it resonated in the halls of Congress, few of whose members had ever seen canyon country.

The Blue Book's title page noted that the report was "sponsored by and prepared under the general supervision of the Bureau of Reclamation, Michael W. Straus, Commissioner." FDR had appointed Straus, who had been an assistant to Secretary of the Interior Harold Ickes, in 1943. Straus broke the mold of reclamation commissioner. He was a rich newspaperman, not an engineer, not from the West, not a self-made man, not a Mormon, and not an old bureau hand. His brother-in-law, Eliot Porter, was the photographer for many beautiful books published by the Sierra Club, including its belated paean to Glen Canyon, *The Place No One Knew*.[6] The Straus-Porter holiday table must have been the scene of some lively discussions, as Straus was the Sierra Club's archenemy.

Harold Ickes presided over the rise of the Bureau of Reclamation during the 1930s and first half of the 1940s. But even as he approved inundating some areas of the West, Ickes arranged to preserve others. The mechanism was the authority of the president to set aside a region as a national monument. In 1908, FDR's fifth cousin Theodore had used the Antiquities Act to protect 800,000 acres of the Grand Canyon. Franklin Roosevelt, only a year after his election, created the Cedar Breaks, Death Valley, and Saguaro national monuments; Joshua Tree and Capitol Reef soon followed. In 1938, Ickes persuaded FDR to enlarge one of the least known and little visited preserves in the country, remote Dinosaur National Monument on the Utah-Colorado border. Bisecting the monument and its spectacular canyons was the Green River, which joined the Colorado several hundred miles downstream, southwest of Moab.

Congress had approved the Antiquities Act in 1906 to prevent looting of archeological sites. But human artifacts are not the only ones of value. Along the Green River, instead of pots, baskets, and spearheads, collectors dug rare dinosaur fossils from the Morrison Formation and shipped them east to natural history museums. In all, some four hundred dinosaur skeletons, including twenty complete specimens, were removed from the quarry near the small town of Vernal, Utah. To prevent further raiding, in 1915 Woodrow Wilson set aside eighty acres of the fossil site as Dinosaur National Monument. But the National Park Service continued to allow the Carnegie Museum of Pittsburgh to extract and remove the dinosaur bones.[7] In 1938, at Ickes's suggestion, FDR expanded Dinosaur to 204,000 acres, making it one of the largest areas protected by the National Park Service. But most of the dinosaur bones were gone and those that remained were encased in solid rock and not easily seen. The monument was difficult to reach and once there, the wild terrain and deep canyons made travel difficult.

Though only a handful of people were aware of it, there was another excellent reason for visiting Dinosaur. A raft trip on the Green River through the spectacular Canyon of Lodore provided a knuckle-whitening, but likely nonfatal, whitewater ride. Here, in one of their first encounters with canyon rapids, Powell and his men had lost one of their boats and one-third of their provisions. In the 1940s and 1950s, Bus Hatch, a carpenter from nearby Vernal, built a river boat and for a small fee would take brave passengers through the canyon. That was the only way to see it other than by inching across the roadless plateau and staring down from high above. There was little other reason to visit Dinosaur and few did. In 1946, the National Park Service counted 7,643 visitors to the monument; in the same year, Grand Canyon, by then a national park, drew nearly half a million.[8] In 1947, only forty intrepid rafters floated the Green in Hatch's river boat.[9]

In 1940, Ickes recommended that FDR accept the by-then several-year-old recommendation of the director of the National Park Service and establish an enormous national monument that would stretch from Lee's Ferry through Glen and Cataract canyons and for scores of miles on either side of the river. At 4.5 million acres, Escalante National Monument would be larger than Yellowstone National Park. Cattlemen used a small percentage of the nearly impassable country for grazing stock, but most of it had nothing to offer but scenery. Nevertheless, the Utah congressional delegation and the few scattered residents of the region, whose ancestors had hacked out a living in one of the most inhospitable parts of the nation, were incensed. They had seen too much of their hard-won state set aside for easterners. The land might not be worth much now, but neither had other western lands before prospectors discovered oil and minerals on them. Even so, Ickes and FDR likely would have gotten their way, as they did in most things, had World War II not intervened and forced the president to turn his attention elsewhere. The giant monument would have included the stretch of the Colorado River near Lee's Ferry and likely would have prevented the bureau from damming Glen Canyon.

When President Bill Clinton arrived at the El Továr Hotel on the South Rim of the Grand Canyon in 1996 to dedicate the 1.7 million-acre Grand Staircase–Escalante National Monument, Glen Canyon was long drowned. Mr. Clinton gave credit for the new monument to Harold L. Ickes: "I'm sorry he never got a chance to see that his dream would become a reality. But I'm very glad that his son and namesake [Harold M. Ickes] is my Deputy Chief of Staff and is here today."[10] The old curmudgeon had gotten his way after all, even if it had taken more than half a century. But one thing had not changed: many Utahans and other westerners remained bitterly opposed to what they saw as usurpation of their lands by executive fiat.

In the run-up to the 2000 presidential election, George W. Bush and Dick Cheney campaigned across the West decrying Clinton's use of presidential decree to establish Grand Staircase–Escalante and vowing to see the decision reversed. Soon after the election, the ultra-conservative Mountain States Legal Foundation sued. Ironically, by then the foundation was forced to sue President George W. Bush, one legal battle that Mr. Bush would have been glad to lose. But in 2004, Judge Dee Benson found that the court had no jurisdiction to decide whether President Clinton had abused his office, and in any case, the claims of the Mountain States Legal Foundation were "without factual or legal support." In late 2005, after the judge's ruling appeared to have laid the matter to rest, the State of Utah appealed on the grounds that although the act does allow the president to designate a monument, he must limit its size.[11] In the summer of 2006, a three-judge panel of the Tenth Circuit Court of Appeals unanimously upheld Judge Benson. The jurists said that in all the monument's vast acreage, not one person could be found who had been harmed by Clinton's decision. Unless someone with standing appeals the case to the Supreme Court, Grand Staircase–Escalante will survive. The struggle represents but one skirmish in a larger battle being waged across the West between the old ways of mining, timber, and grazing and the new revenue sources of recreation and ecotourism.

In 1940, Congress directed the bureau to conduct "studies and investigations . . . for the formulation of a comprehensive plan for the utilization of waters of the Colorado River system for irrigation, electric power, and other purposes." Appearing six years later, the Blue Book described so many projects that it was a rare farmer in the four states who did not appear to have a stake in at least one. Yet as Harry Truman's secretary of the Interior Oscar Chapman acknowledged in the foreword to the Blue Book, there was not nearly enough water for all of the 134 projects. In hindsight, this may have been an ideal bureaucratic strategy, leading many more to support the overall program than could possibly benefit in the end.

By 1946, the Bureau of Reclamation had completely reversed its pre-Hoover Dam image and was riding high. Its annual funding would soon grow to account for one-third of the entire budget of the Department of the Interior. The bureau had undertaken huge projects on the Columbia, the Missouri, and the Sacramento. But to fuel the juggernaut, to occupy its large staff and justify its funding, the bureau required a constant supply of new dam sites.

The need and the opportunity inspired the authors of the Blue Book to eloquence:

> Yesterday the Colorado River was a natural menace. Unharnessed it tore through deserts, flooded fields and ravaged villages. It drained the water from the mountains and plains, rushed it through sun-baked thirsty lands, and dumped it into the Pacific Ocean—a treasure lost forever. Man was on the defensive. He sat helplessly by to watch the Colorado waste itself, or attempted in vain to halt its destruction.
>
> Today this mighty river is recognized as a national resource. It is a life giver, a power producer, a great constructive force. Although only partly harnessed by Boulder dam [to be renamed Hoover Dam] and other ingenious structures, the Colorado River is doing a gigantic job. Its water is providing opportunities for many new homes and for the growing of crops that help to feed this nation and the world. Its power is lighting homes and cities and turning the wheels of industry. Its destructive floods are being reduced. Its muddy waters are being cleared for irrigation and other uses.
>
> Tomorrow the Colorado River will be utilized to the last drop. . . . Here is a job so great in its possibilities that only a nation of free people have the vision to know that it can be done and that it must be done. The Colorado River is their heritage.[12]

The Blue Book listed a number of potential dam sites, but being so far upstream, few were deep enough to serve as true cash-register dams. Even cursory financial scrutiny would show that the high-altitude, short-season irrigation projects were bound to lose money. How could the bureau make the project as a whole show a positive benefit-cost ratio and thus justify funding?

The answer was "river-basin accounting." The bureau applied the novel method first on a small project in Wyoming and then in the giant Missouri River Plan, one of the largest river development schemes ever, so large that the bureau had to invite its archenemy, the Army Corps of Engineers, to share in its construction. By lumping cash-register dams and money-losing irrigation projects into one giant account, the whole set could be made to come out in the black. Outside of government, the idea would have been ludicrous and have quickly led to bankruptcy. Take the six divisions of General Motors in the early 1950s: Buick, Cadillac, Chevrolet, GMC Truck, Oldsmobile, and Pontiac. Suppose that three of the six consistently lost so much money that the profit made by the other three was barely enough to keep the parent company afloat. What would GM's shareholders insist that the company do? Turn the money-losing brands around, sell them, or close them down. Had GM operated instead like the Bureau of Reclamation, the money-making divisions would have gone on subsidizing the money-losers until, having no taxpayers to bail it out, the parent company went broke.

The total construction cost of the projects listed in the Blue Book came to $2.185 billion in 1940 dollars. To repay that amount at 3 percent interest over thirty years, the standard rate and duration the bureau used in its projections, would cost $85 million annually. Add the annual operating expense and the total yearly cost came to $108 million.

The bureau estimated annual benefits at $65 million from newly irrigated crops, $72 million from the sale of 1,622,000 kilowatts of electricity, $500,000 from water sales, and $1 million from flood control—that is, the estimated value of unidentified properties not lost to future floods. With benefits totaling $138.5 million, the annual benefit-cost ratio came to 1:28. For each dollar spent, the Colorado River Storage Project (or CRSP), as the plan was now called, would return $1.28. "There are . . . numerous other intangible benefits," the report claimed, that are "none the less real." It concluded, "Studies of a general nature show that a program for complete river development would be completely justified. Direct evaluated benefits would exceed the costs even though many public benefits are not considered."[13] That there might also be numerous other costs, tangible and intangible, was not mentioned. But it did not matter. As we will see, the Bureau of Reclamation either ignored the results of its benefit-cost analyses or finagled them until they came out sufficiently positive.

But who would actually pay for the construction and the annual cost of operation of the various CRSP components? Not the irrigators who were to be its main beneficiaries. According to the Blue Book, they could pay back no more than $8 million of the $65 million in annual sales. The chief hydrologist of the bureau told Congress that even that fraction was too optimistic: "The [upper basin] farmers can't pay a dime, not one dime," he said.[14] The bulk of the revenue, the $72 million from power sales, would have to make up the difference. Unlike at Hoover Dam, where consumer payments for electricity covered the cost of dam construction, in the upper basin electricity users would in effect pay farmers to irrigate.

Remember that the Colorado River Compact did not specify how the states within each basin were to divide their 7.5 MAF annual allotment. Until that was settled, any attempt to fund Blue Book projects would only lead to internecine bickering as each state tried to get its share and then some. Chapman shrewdly required that before Congress approved any upper basin projects, "within the limits of the general allocation of water between the upper basin and lower basin states set out in the Colorado River Compact, the Colorado River Basin States [must] agree on suballocations of water to the individual states."[15]

With no Californians to threaten them, Colorado, New Mexico, Utah, and Wyoming quickly agreed. By the late 1940s, no one could deny that the Colorado River carried less water than the compact had allocated. Instead of allotting each state a specific volume, the four decided to use percentages: 51.75 percent to Colorado, 11.25 percent to New Mexico, 23 percent to Utah, and 14 percent to Wyoming.[16] In recognition that Arizona contributes a small amount of water to the upper basin, the Upper Colorado River Basin Compact allocated the state 50,000 acre-feet. Representative Wayne Aspinall from Colorado's western slope near Grand Junction introduced legislation in 1948 to ratify the upper basin compact and it quickly passed. Before long, Aspinall and Arizona senator Carl Hayden would become the two most powerful people in the country on western water issues.

As the 1952 presidential election approached, the Republican candidate, military hero Dwight D. Eisenhower, was the most popular man in the nation. Ike struck fear into Mike Straus and the bureau by announcing during the campaign that his policy toward federal reclamation would be "no new starts." The Republican Platform explained why:

> We vigorously oppose the efforts of this [the Truman] national administration, in California and elsewhere, to undermine State control over water use, to acquire paramount water rights without just compensation, and to establish all-powerful Federal socialistic valley authorities.[17]

Douglas McKay, an Oregon car dealer whom Ike had tapped as his Interior secretary, announced before the inauguration that one of his first acts would be to "fire Mike Straus."[18] These threats made the upper basin states understandably nervous. Just when they were on the verge of getting the dams, reservoirs, and irrigation projects they had long sought, Ike's likely election and rapacious dam-builder Straus's likely departure promised to dash their dreams. For the Blue Book and the upper basin, time was fast running out.

Straus's strategy was to get the CRSP far enough along so that political pressure from the upper basin states would make it difficult for the incoming administration to reverse course. From the long list in the Blue Book, Straus chose four hydropower dams and eleven other reclamation projects. One of the cash-register dams would be a 529-foot arch that would

Figure 9. Echo Park, looking downstream. Steamboat Rock is on the right; from the left, the Yampa quietly joins the Green. (U.S. Geological Survey)

impound 6.4 MAF and generate 200,000 kilowatts. It would rise in a place hardly anyone had heard of called Echo Park, at the confluence of the Green and Yampa rivers. The site lay squarely within the supposedly protected confines of Dinosaur National Monument. That fazed the Bureau of Reclamation not in the least. It knew that when FDR enlarged the monument in 1938, he had explicitly allowed the possibility of power dams. The largest of the four cash-register dams in the CRSP, to be only a few feet lower than Hoover, would go up at the foot of Glen Canyon, which lacked protection and which hardly anyone on either side of the coming debate had even seen or knew the slightest thing about.

Arizona senator Barry Goldwater was a rare exception. In 1940, he had rafted both Echo Park and Glen Canyon with pioneer river man Norm Nevils, making him likely the only member of Congress who had seen both. The soon-to-be Republican candidate for president voted for the CRSP dams, but as he neared retirement said that if he could recast one vote from his Senate career, it would have been the vote that doomed Glen Canyon.[19]

The California delegation, on the other hand, did not need to know anything about either Echo Park or Glen Canyon to oppose damming them. Its representatives could be counted on to object to any project that might prevent water from flowing unchecked to the Golden State. But the bureau was unconcerned. It had hardly ever been challenged, much less defeated. After the usual tiresome machinations from the Californians and a little horse-trading, the CRSP would surely sail through Congress unscathed.

True, figured the bureau, a handful of people calling themselves conservationists, the progeny of John Muir and his ilk, would likely protest, but they were poorly organized, had no influence on public policy, and had never been to either Echo Park or Glen Canyon. One exception was engineer Walter Huber, the former president of a group of elite alpinists calling itself the Sierra Club. But Huber reported that Echo Park was not worth saving—"just canyons and sagebrush . . . nothing to be concerned about."[20] And the Sierra Club's new executive director certainly posed no threat to the powerful bureau. An amateur lepidopterist, his math skills did not extend much beyond those of grade school, where he had been nicknamed the "toothless boob."[21] He had never testified before Congress and earlier, as a member of the Sierra Club board, had voted to dam the Grand Canyon.

The New Deal and World War II had made the Bureau of Reclamation a national priority and one of the most powerful government agencies in this or any country. Its rise had positioned the bureau to get everything it wanted on the upper river, including high dams in Dinosaur National Monument and Glen Canyon. The juggernaut of federal reclamation gathered itself, ready to quell the menacing Colorado River and turn it into a series of innocuous millponds.

NINE

Shall We Let Them Ruin Our National Parks?

CONGRESS ESTABLISHED THE National Park Service in 1916 and gave the agency the mandate to "conserve the scenery and the natural and historic objects and the wildlife therein and to provide for the enjoyment of the same in such manner and by such means as will leave them unimpaired for the enjoyment of future generations."[1] According to the Oxford English Dictionary, *to conserve* "suggests keeping sound and unimpaired and implies the use of means to prevent unnecessary or excessive change, loss, or depletion."[2] Thus the park service must manage the difficult balancing act of allowing the present generation to enjoy wilderness, yet in such a way as not to deprive future generations of the same opportunity.

In the original Latin, *reclamare* ("to reclaim") was "to cry out against, contradict."[3] But one does not have to turn to etymology to understand that the Bureau of Reclamation and the National Park Service have opposite missions. By definition, the park service conserves wilderness, whereas the bureau alters it beyond recognition, replacing wild canyons with placid reservoirs.

—

As the funding and prowess of the bureau rose during the 1930s and 1940s, the park service found itself pulled along behind. After the bureau's first big success at Hoover Dam, Congress established the Lake Mead National Recreation Area and asked the park service to administer it. As the bureau went on to plan other reservoirs, the government intended that the park service would manage recreation areas at those sites as well. Eventually the bureau was bound to propose a dam at a site that already had protection as a national park or monument. Then the conflicting missions of the two agencies were bound to collide. The conflict led to the seminal conservation battle of the twentieth century, launched the modern environmental movement, and led directly to the raft of environmental legislation of the coming two decades. The battleground was another spot that even the conservationists had barely heard of: Echo Park.

After leaving the Canyon of Lodore on the Colorado-Utah border, the Green River makes a sweeping, 180-degree bend around a steep-walled slab of rock called Steamboat Mountain. Just before the bend, the Yampa River, having descended its own spectacular canyons, enters quietly from the east. The two rivers have created not only some of the most striking scenery in the West but also echoes that Powell's crew said they heard repeated as many as ten times. The site was enough to make any engineer salivate. A single dam just below the confluence would produce a Y-shaped, two-for-one reservoir that would back water up both the Green and the Yampa and impound twice the volume for the money.

The bureau wasted no time demonstrating whose mission it thought should prevail. In 1939, only one year after FDR expanded Dinosaur National Monument and without clearing it with the park service, the bureau built a rough, thirteen-mile-long road from the rim down to the floor of Echo Park and began to conduct surveys and drill test holes. Two years later, evidently having decided that opposition was futile, the park service agreed that the two agencies would find a way for Dinosaur National Monument to "serve the purposes of both bureaus," in retrospect, an impossible contradiction.[4]

World War II further emasculated the park service, relegating it to a "non-essential function" headquartered in Chicago, not Washington. Its budget shrank from $21 million in 1940 to $5 million in 1943 and its staff size fell in proportion. The park service had become a nearly irrelevant Bambi dodging Godzillas at war.

By 1944, National Park Service director Newton Drury decided that rather than allowing a dam in a national monument, the president would better downgrade Dinosaur to a national recreation area, which carried fewer protections.[5] The strategy would sacrifice Echo Park to preserve the inviolability of the national park system—like amputating a gangrenous limb to save a life. But if the disease spread, doctors might also have to remove other limbs and even vital organs. The bureau was known to be planning dams in national parks across the country—who could guarantee that with a precedent at Echo Park, Congress might not revoke the status of one or more of those parks as well?

World War II put the bureau's plans on hold. By the time the Blue Book appeared in 1946, the war was over. No longer was it self-evident that hydropower should trump recreation. Indeed, many American families, their men and women home safely, with Detroit returning to automobile production, with gasoline once again plentiful, wanted to see the USA by car, including its fabled western parks and monuments. The section of the Blue Book written by the postwar National Park Service showed that its backbone had finally begun to stiffen:

> The canyons of the Green and Yampa Rivers . . . possess great importance . . . as an introduction to the geology and scenery of the West [and are] of national significance . . . justif[ying their] existence as a unit of the National Park System. The Echo Park and Split Mountain reservoirs[6] . . . would cover . . . a number of notable geological formations, would reduce the visible height of canyon walls . . . and would substitute long bodies of still water . . . for the natural streams and vegetations in canyon bottoms.
>
> Before changes . . . are authorized in order to recognize water control as the principal consideration in administering the unit, it should have been clearly and certainly shown that it would be in the greater national interest to develop the area for such use than to retain it in its natural state for its geologic, scenic and associated values and for the enjoyment of them by the Nation.[7]

To jump ahead of our story in order to show the continuing evolution of the park service's attitude, in 1954, as the debate over the dams in Dinosaur began in Congress, one anti-dam member had read into the record a report that the National Park Service had issued a year earlier on "encroachments" onto park service lands. No one could mistake how far the attitude of the park service had shifted. "The greatest peril to the parks from dam proposals," the report accused, "comes from the plans and programs of the governmental dam-building agencies themselves and the pressures which their activities generate in the various sections of the country."[8]

In the immediate postwar years, the bureau roamed the West, touting the advantages of dams to whomever would listen, all the while ignoring the park service. The more determined the bureau became to dam Echo Park, the more the mettle of park director Drury rose. In an Interior Department survey of recreational resources afforded by the Colorado River, he retaliated. A dam in Echo Park, Drury wrote, "would be totally alien to the geology and landscape of the monument . . . a lamentable intrusion . . . power lines could hardly be constructed without injury to geological and scenic values. . . . Even more unfortunate, would be submersion of the floor of Echo Park."[9]

Drury pointed out to Secretary of the Interior Oscar Chapman that though the Blue Book had listed many dam sites, only one had become indispensable: the only one located inside a national monument. Why could the bureau not substitute a site outside the monument, or some combination of sites, for the one in Echo Park, and thus avoid the collision of the two opposite missions? Commissioner Straus replied with the argument the bureau would use repeatedly in the fight: any site or combination that did not include the Echo Park dam would cause more evaporation and waste precious water.[10]

With the bickering between two of his key agencies becoming an embarrassment and delaying the entire CRSP, Chapman decided to air the matter at a public hearing. On 3 April 1950, he opened the one-day hearing, noting that his decision would rest on what he perceived to be "the greatest good for the greatest number of people."[11] Looking down on the attendees from the wall of the cavernous auditorium of the Department of the Interior was a bas-relief of John Wesley Powell.[12] Opponents and proponents of the Echo Park dam alike no doubt believed that the Major, up in the heavens of geology, was on their side. But the wise Washington hand kept his counsel.

Ready to testify were conservationists, Senator Arthur Watkins of Utah and three other western senators, five members of the House, Bureau of Reclamation officials, and representatives of business interests from the upper basin. Bureau engineers had the floor first. They cited the need to regulate the Colorado and to store water so that the upper basin could meet the terms of the Colorado River Compact in dry years. Nothing less than 48 MAF of storage would do, a bureau expert testified— even though that was at least three times the annual flow of the Colorado

River. Following the party line that Straus had laid out, the overriding reason for damming Echo Park was to minimize evaporation. It was a good strategy, for unlike such vague concepts as the value of scenery and the greatest good, evaporation can be quantified. Using criteria other than evaporation, the bureau could first reduce the number of possible dam sites in the upper basin. Then it could use the trade-off between evaporation and storage to choose the best from among the remaining sites. The bureau's assistant director for project planning testified that "any group of reservoirs which does not include Echo Park and Split Mountain can meet the objectives heretofore outlined only at the cost of increased evaporation loss, less annual revenues, and higher unit power costs." The next best combination of sites, he explained, would lose a minimum of 350,000 acre-feet more water annually through evaporation, enough to supply a city of 1.5 million. In 1950, that was three times the size of metropolitan Denver.[13]

National Park Service director Drury spoke against the dam. So did Ira Gabrielson, the influential former director of the U.S. Fish and Wildlife Service and, at the time of the hearing, president of the Wildlife Management Institute. "Flowing water is an essential part of an exhibit of the Colorado River," Gabrielson claimed. "Conservationists do not subscribe . . . to this specious argument that the 'stinking debris' left by a fluctuating reservoir adds to the beauty of any natural scenery." Gabrielson summed up, "Forever is a long time, but what is lost . . . by these huge impoundments is lost forever."[14] Kenneth Morrison of the National Audubon Society foresaw a bleak future: "No one has ever been able to place a dollar sign on wilderness values. Who can say what the value of an unspoiled Dinosaur National Monument will be in an era when the face of the nation has been almost completely irrigated, drained, and dammed?" He noted that "New Yorkers could derive a tremendous amount of tax revenues from the acreage of Central Park, if it were to be filled in with skyscrapers."[15]

Morrison and the other conservationist leaders knew that the bureau and Army Corps of Engineers planned dams in several other protected areas. Dams in the Grand Canyon would back water up into both its national monument and national park. Glacier View Dam on the Flathead River would inundate 20,000 acres of Montana's Glacier National Park. A high dam would drown King's Canyon National Park in California; another on Kentucky's Green River would flood the underground Echo River in Mammoth Cave National Park and extirpate its unique sightless fish and other specialized creatures. Forty miles of the historic

Chesapeake and Ohio Canal on the Potomac would disappear under federal reservoirs.[16] And these were merely the sites in national parks and monuments that the bureau and the corps had gotten around to by the end of the 1940s. If the Bureau of Reclamation could dam Dinosaur National Monument, a precedent might be established that would leave every national park and monument vulnerable and render congressional protection meaningless.

One who spoke against the dam had a particularly impressive name and set of credentials. General Ulysses S. Grant III had served for forty-three years with the Bureau of Reclamation's rival, the Army Corps of Engineers. His experience had included six years as U.S. district engineer for the Sacramento and San Joaquin river system in California. Grant had built dams; now, speaking as president of the American Planning and Civic Association, he challenged the bureau's claims that a dam in Echo Park was indispensable. He set the stage by noting that he was "the grandson of the president who established Yellowstone National Park and who started this whole national park system."[17]

In spite the bureau's claims, Grant said, the Echo Park and Split Mountains dams were not essential. Instead, he asserted, "A slight change in order and a substitution . . . of other projects for these two projects . . . will answer the purpose just about as well." He challenged the argument that dams and reservoirs provide "better and more recreational facilities." Instead, those opportunities would be the same as on "all the other reservoirs." In contrast, "The natural features of Dinosaur National Monument are . . . unique."[18]

In a September 1951 article in the official organ of the American Planning and Civic Association, published after Chapman's decision on the Dinosaur dams but before Congress took up the CRSP, General Grant had room to lay out his case. The bureau's assertion that without a dam at Echo Park, evaporation would cost an additional 350,000 acre-feet, he said, was "just a guess, a nice big round figure to help the sales argument." Foreshadowing the continuation of the debate four years later, he claimed that the estimate was "apparently not even correct based on the Bureau's own method of estimating." The agency, he accused, "had made a simple calculation mistake."[19]

As dam proponents had pledged their love of wilderness, so dam opponents acknowledged the vital importance of water. With both sides agreeing that the nation must have both dams and wilderness, compromise was in order. But on what ground could the conservationists compromise? Where could they draw the line between wild areas that were off limits and those that were not? The only nonarbitrary place was to recognize that Congress and the president had already set aside certain areas of the country as national parks or monuments. The founding language of the National Parks Act declared that they are to be conserved and left unimpaired. Here was the only logical and legal platform on which the conservationists could make their stand.

On 27 June 1950, Secretary Chapman came down in favor of the Echo Park Dam.[20] His letter noted that the presidential order expanding Dinosaur National Monument had contemplated its use for water projects and therefore that his decision "will not provide a precedent dangerous to other reserved areas." But as historian Mark Harvey discovered, Chapman had a more important, though secret, reason for his decision. On 29 August 1949, eight months before the hearing, the Soviet Union had exploded a 22-kiloton atomic bomb, shocking and terrifying the West and launching the Cold War nuclear arms race. President Truman responded by authorizing the building of the hydrogen bomb, to be tested in Utah and Nevada. As Harvey points out, what happens behind the scenes in Washington is usually more important than testimony in a hearing room. Representatives of the Atomic Energy Commission told Chapman in secret that they would need more electrical power than was available in Utah, making hydropower dams a matter of national security—even of national survival.[21] In an oral interview more than twenty years after the Echo Park battle, Chapman said, "I was opposed to seeing that dam built in that park . . . the dam would serve just as well if built 50 miles downstream [thus beyond the monument boundary]."[22] Then as today, national security issues trumped all others. Evidently they led Chapman to subordinate his conservationist instincts.

The conservationists had lost the first skirmish, but Chapman's decision did nothing more than approve continued planning—it did not set aside money for the hugely expensive CRSP. Congress itself would have to authorize the program and then appropriate funding. There was still a long way to go before a high dam would rise in Echo Park.

At the time Chapman announced his decision in late June 1950, hardly anyone outside of government and the uncoordinated conservationist groups knew that a Dinosaur National Monument existed, much less that the bureau wanted to dam its most beautiful stretches. The country at large got wind of the controversy in article published on 20 July 1950 in the immensely popular *Saturday Evening Post,* several weeks too late to influence Chapman but in time to alert the public. The author was Bernard DeVoto, a native of Ogden, Utah; Harvard graduate and part-time professor of history there; member of the National Parks Advisory Board; and author of a regular column in *Harper's Magazine* dubbed "The Easy Chair." From the catbird seat of his column, the curmudgeon had already taken on such western topics as the absurdly cheap grazing leases awarded to large cattle operations by the Bureau of Land Management, one of many "land grabs" in the West that DeVoto had labeled "the plundered province." The attitude of his native westerners toward the government, DeVoto said, was "get out and give us more money." He warned his readers, "This is your land we are talking about."[23] DeVoto titled his *Post* article, "Shall We Let Them Ruin Our National Parks?"[24] Getting to the point immediately, his subtitle read, "Do you want these wild splendors kept intact for your kids to see? Then watch out for the Army Engineers and the Bureau of Reclamation—because right where the scenery is, that's where they want to build dams." He explained:

> Echo Park Dam would back water so far that throughout the whole extent of Lodore Canyon the Green River, the tempestuous, pulse-stirring river of John Wesley Powell, would become a mere millpond. The same would happen to Yampa Canyon. Throughout both canyons the deep artificial lakes would engulf the magnificent scenery, would reduce by from a fifth to a third the height of the precipitous walls, and would fearfully degrade the great vistas. Echo Park and its magnificent rock formations would be submerged. Dinosaur National Monument as a scenic spectacle would cease to exist.[25]

Dinosaur was important "as wilderness that is preserved intact . . . for the field study . . . of the balances of Nature, the web of life, the interrelationships of species, massive problems of ecology—presently it will not be possible to study such matters anywhere else." In a democracy, the voters ought to have a voice, DeVoto argued; yet, he continued, "No one has asked the American people whether they want their sovereign rights, and those of their descendants, in their own publicly reserved beauty spots wiped out. No one can doubt that the public, if told all the facts and allowed to express its will, would vote to preserve the parks from any alteration now or in the future." Instead, "by the time a project is laid before Congress it has already been decided upon, the local interests have been organized and the Western senators and representatives—one of the most powerful blocs in Congress—have been lined up. Within the West there is severe infighting for the allocation of projects, but when it comes to projects to be allocated, there are neither state nor party lines: there is only a solid West." The National Park Service itself, DeVoto believed, was "in danger of being subverted by engineering construction."

The widely circulated *Reader's Digest* printed a shorter version of DeVoto's *Post* article, and other magazines followed suit. His fellow westerners accused him of treason and complained to the editor of the *Saturday Evening Post,* which never again published DeVoto.[26] The *Deseret News* and the *Denver Post* took him to task, prompting DeVoto to reply: "The National parks and monuments happen not to be *your* scenery. They are *our* scenery. They do not belong to Colorado or to the West, they belong to the people of the United States." He warned westerners not to "snoot [easterners] too loudly or obnoxiously. You might make them so mad that they would stop paying for your water development."[27]

The negative publicity enraged former newspaperman Mike Straus. In a speech at the dedication of one of the pumping stations of the Colorado–Big Thompson Project, he let the DeVotos of the world have it: "From their air-conditioned caves overlooking the undeveloped wilderness areas of Central Park in New York, Lincoln Park in Chicago, and Boston Commons in the adopted city of a transplanted westerner who has a tendency to forget his heritage, these self-appointed guardians . . . contend . . . that the highest use for your area and resources is as a museum and cemetery for dinosaur bones."[28] This was sheer propaganda: Straus knew as well as anyone that the water from a reservoir in Echo Park would not reach the dinosaur fossils—but he did know that it would drown the scenic canyons.

Part of Straus's anger rose from his suspicion that DeVoto and Drury were in cahoots—could the park service director and his staff even have collaborated on DeVoto's article? In the last act of their marriage of convenience, like an angry spouse demanding to see the secret love letters tied up in a ribbon of blue, Straus demanded that Drury show him copies of all his correspondence with DeVoto.[29] Straus may have had reason not to trust Drury. According to Chapman, the park service director had repeatedly denied having signed an agreement to permit the bureau to dam Echo Park—even while Chapman had in his pocket a copy bearing Drury's signature.[30]

In December 1950, Chapman proposed to Drury that he leave the National Park Service and become a special assistant to the secretary of the Interior, with a portfolio that would include "conservation matters."[31] In fact the offer was nothing more than a way to neutralize Drury by giving

Figure 10. Bernard DeVoto, crack writer and crack shot, 1936 (Courtesy of Mark DeVoto)

him nothing to do. When Chapman announced he would appoint Drury's successor on 1 April, the park service director wisely resigned and went home to California to head the state's Division of Beaches and Parks. No love was lost in the other direction. In an interview two decades later, Drury said that Chapman was "utterly impotent in the hands of his subordinates. He was very much in the position of the mahout who rides the elephant and thinks he's guiding it but really is being carried along. The great Bureau of Reclamation was like the state of Prussia in the German empire, where everything was weighted in its favor."[32]

Naturalist John Muir founded the Sierra Club in 1892 and served as its first president. The club's purpose, in Muir's words, was to "do something for wildness and make the mountains glad."[33] With Muir in the lead, the Sierra Club fought a long battle to prevent the city of San Francisco from damming Hetch Hetchy Valley, a sister canyon just to the north of Yosemite and in the eyes of many, an equally beautiful one. In the end the club failed and Hetch Hetchy disappeared. But perhaps not forever—Environmental Defense and other organizations today are searching for another source of water for San Francisco in order to allow Hetch Hetchy to be restored. The battle for Echo Park gave the Sierra Club a second chance to prevent a dam in another beautiful canyon that the president had declared should remain unimpaired. But first the Sierra Club had some of its own recent history to overcome.

The two most costly projects in the Blue Book were dams at Marble Canyon–Kanab Creek and another at Bridge Canyon, both squarely within the Grand Canyon. A dam at the Marble Canyon site would back water up fifty-three miles, thirteen of them within Grand Canyon National Park. A dam downstream of the park at Bridge Canyon, at river mile 235 just above Separation Canyon, would flood ninety-three miles, forty of them in Grand Canyon National Monument. To the horror of conservationists, just as FDR's expansion of Dinosaur National Monument had allowed for the possibility of hydropower dams, so Congress had long before approved in principle the damming of the Grand Canyon.

When Congress set aside Grand Canyon National Park in 1919, Representative Carl Hayden of Arizona saw to it that no mining claims or valuable timberlands fell within the designated park boundaries. The wily Hayden also made sure the act stated, "Whenever consistent with the primary purposes of said park, the Secretary of the Interior is authorized to permit the utilization of areas therein which may be necessary for the development and maintenance of government reclamation projects."[34] In other words, whenever the Bureau of Reclamation wanted to dam the Grand Canyon, it already had the authority.

Now the time had come, prompting the board of directors of the Sierra Club to respond. At a special meeting on 12 November 1949 in Los Angeles, the board unanimously passed a motion that expressed the policy of the Sierra Club toward dams in the Grand Canyon:

> The reservoir created by this [Bridge Canyon] dam will submerge portions of Grand Canyon National Park and Grand Canyon National Monument. In the event that the appropriate authorities determine that the construction of such dam is economically sound, thorough consideration should be given to minimizing the impact of such dam and reservoir on the scenic and inspirational features preserved for public use in the creation of Grand Canyon National Park and Monument. Legislative action is also necessary to insure that the heart of Grand Canyon will not be invaded by future dams and diversions without express permission of Congress. To accomplish these purposes the Sierra Club recommends:
>
> 1. The construction of Bridge Canyon Dam should not be authorized unless necessary prior action has been taken to insure the construction of Glen Canyon and Coconino Dams [the Blue Book also proposed a dam at Coconino in the Grand Canyon above Bridge Canyon] or equivalent dams on the main stem of the Colorado and on the Little Colorado to prevent siltation of the Bridge Canyon Reservoir. Without such prior construction engineering estimates indicate that the upper forty miles of Bridge Canyon Reservoir will be filled with silt and rendered unusable for public recreation and inspiration in a period of three and one-half years [four other recommendations followed].[35]

The club might have opposed even one dam in the most sublime scenery in the nation, site of both a national park and a national monument. Instead its board acquiesced to the bureau and moved on to how, rather than whether, the bureau ought to build dams in the Grand Canyon. It endorsed a dam at Bridge Canyon *as long as* the bureau built two other dams upstream, one at Glen Canyon. Not one dam, but three. As testament to the hegemony of the Bureau of Reclamation at mid-century, the Sierra Club, the seminal conservation organization, evidently believed it had no choice but to surrender the Grand Canyon without firing a shot. But the club's subservience was soon to end.

In 1950, before the bill to authorize the CRSP had been introduced, now-Senator Carl Hayden sponsored one of his own to authorize the Bridge Canyon dam. The dam's main purpose would have been to generate power to drive pumps to lift Colorado River water up to the cities of central Arizona. The bill made it through the Senate but foundered in the more divisive House, where the pugnacious Arizona and California delegations continued to turn any water issue into a fight to the finish.

Before 1951, one could not simply join the Sierra Club—one had to be endorsed by a current member. The club's purpose was regional and limited: "to explore, enjoy, and render accessible the mountain regions of the Pacific Coast." That same year the club board opened its membership and changed its mission to read: "explore, enjoy, and protect the Sierra Nevada and other scenic resources of the United States."[36] The substitution of "protect" for "render accessible" and the expansion of the Sierra Club's purview to the entire country showed that, already ruing its acquiescence to dams in the Grand Canyon, the club was ready to fight a dam in Echo Park. To lead that fight and to put its new mission into practice, the Sierra Club needed someone who could hold his own against the powerful Bureau of Reclamation, something that no organization or individual had succeeded in doing over the past fifty years. To take on Goliath in the Valley of the Colorado, conservationists needed a David.

TEN

We Want to Be Dammed

THOUGH THE BUREAU OF Reclamation could have taken comfort from the Sierra Club's recent support for dams in the Grand Canyon, the biography of its new executive director should have sounded an alarm. During World War II, butterfly collector David Brower had served with the elite Tenth Mountain Division of white-camouflaged Alpine fighters and won the Bronze Star. He recorded over seventy first ascents of peaks in the Sierras and elsewhere, including Shiprock, the classic New Mexico volcanic neck. Brower, the college dropout, went on to serve as editor of the Sierra Club's magazine as well as at the University of California Press. The tireless, eloquent, charismatic Brower would turn out to be the bureau's most formidable opponent during the twentieth century.

When the battle for Echo Park got under way, Brower and the other conservationists lacked the arsenal that Congress would provide in the raft of environmental legislation of the 1960s and 1970s. The requirement of the National Environmental Policy Act of 1969 (NEPA) that an environmental impact statement precede any large project would by itself have barred dams in both Echo Park and Glen Canyon. Congressman Wayne Aspinall saw it that way: "We got the CRSP approved just in time," he said. "Today we could never get it authorized—particularly if it included Glen Canyon Dam."[1] Floyd Dominy, the archetypal reclamation commissioner, would make the same point decades later: "Why today, we'd have

Figure 11.   David Brower ascending Shiprock, 1939 (Photo by John Dyer; courtesy of Earth Island Institute)

spent six months trying to find out where to put the toilets for the rock scalers. They've got so many lint picking organizations lookin' over the shoulder now that there's no way you can get a project approved."[2]

But without NEPA, the Endangered Species Act, the Clean Water Act, and the rest of the environmental legislation, the conservationists had to invent their weapons as they went along. Brower began by bringing Sierra Club members to Dinosaur National Monument to see for themselves what the dam would destroy, hoping they would spread the word. He arranged for Bus Hatch to guide small groups of club members on raft trips down the Green River through Lodore, Echo Park, and Split Mountain. To introduce Dinosaur to a wider audience, Brower toured with a color film called *Wilderness River Trail.* He placed articles opposing the dam in the *Los Angeles Times,* the *San Francisco Chronicle,* and, of special importance given its wide audience, *National Geographic Magazine.* Brower even entered the lion's den—the chamber of commerce for Vernal, Utah—to tell the Mormon businessmen that they should "leave these canyons the way they are, provide reasonable access to them, schedule a variety of trips on the river, and the world will beat a path to your door."[3] The down-to-earth Vernalites were not inclined to agree. When the curator of the Utah Field House of Natural History in Vernal appeared before the House subcommittee that was considering the CRSP, he told how he had responded when a member of a visiting Sierra Club group "spilled the beans." The slippery rafters had come to Vernal one summer not to enjoy nature as they pretended, but "to save Dinosaur National Monument for you people so they won't build those dams in there." As the curator recalled his response, he had told the outsiders, "That's certainly very nice of you and I'm sure you are prompted by the very best of motives, but did it ever occur to you that we might not want to be saved? As it so happens, we don't. We want to be dammed."[4]

—

The battle over Echo Park began in the House Subcommittee on Irrigation and Reclamation in January 1954, almost four years after the Chapman hearing. In the meantime, the bureau had pared the long list of projects in the Blue Book down to four cash-register dams, including Echo Park and Glen Canyon, and eleven other irrigation projects. Its top engineers and most persuasive spokesmen were briefed and ready, confident that no one could challenge them—after all, no one ever had. They could take comfort from the presence in the hearing room of Representative Wayne Aspinall, from Palisade, Colorado, on the state's western slope, only a few score miles from Dinosaur National Monument. Aspinall had been in office only since 1948, but already had moved well up in seniority on the parent House Committee on Interior and Insular Affairs. Soon he would chair the committee. The engraving on his Colorado gravestone is headed simply, "Mr. Chairman." All by himself, Wayne Aspinall anchored one side of the Concrete Pyramid. He knew what was at stake at Echo Park. Losing the dam "would give conservationists a tool they will use against us for a hundred years," he warned.[5] Brower would later say, "We have seen dream after dream dashed on the stony continents of Wayne Aspinall."[6] Central casting could not have come up with better-suited opponents than the lean, patrician alpinist and the short, bespectacled, bulldog-like former schoolteacher.

For the first six of the ten days of hearings, Bureau of Reclamation engineers and representatives of the pro-dam forces held the floor. They knew that the conservationists would attack at the weakest point: the proposal to build a dam inside a national monument. The bureau used the same strategy it had at the Chapman hearing four years earlier: make the Echo Park dam indispensable to the entire CRSP. For their part, the conservationists knew they were on the defensive. Eisenhower's secretary of the Interior, though only briefly in office, had already earned the nickname "Giveaway McKay" for his failure to protect parks and monuments.[7]

Ralph Tudor, undersecretary of the Interior, began the bureau's presentation. Tudor's credentials were impeccable: a graduate of West Point, he had served as an engineer with the U.S. Army, headed up a profitable private company, and helped design the San Francisco Bay Bridge.[8] The year before the hearing, Tudor had rafted the Green. He had reported to Douglas McKay that "the alteration will be substantial," adding that, "if conflicting interests did not exist," he would prefer to see Dinosaur "remain in its natural state." But of course, where water in the West is concerned, conflicting interests inevitably do exist and up to the time of the hearings, those of the water establishment had always trumped those of the conservationists. Later in his testimony Tudor said that "the choice is simply one of altering Dinosaur National Monument without destroying it." When his turn came, David Brower rejoined that if a dam would only alter Echo Park, then "a dam from El Capitan to Bridal Veil Fall would not destroy Yosemite, but just alter it."[9]

Tudor testified that even the best combination of sites that did not include a dam in Echo Park would evaporate 165,000 acre-feet more water annually.

But wait a moment: at the Chapman hearing four years earlier, bureau experts had estimated the additional evaporation at 350,000 acre-feet. At the time, Grant had accused the bureau of making a calculation error. Evidently it had, because now, without explanation, the agency cut its estimate of evaporative loss roughly in half.

In preparing his testimony, Brower sought the advice of one of the foremost hydrologists in the country, Luna Leopold, son of famed pioneer environmentalist Aldo Leopold. Luna Leopold warned Brower that the bureau was far more likely to be right on water matters than any consultant the Sierra Club could hire—and by implication, Brower himself, who had not the background to be a consultant. To challenge the bureau on the ground of its greatest expertise would be the "height of folly," said Leopold.[10] Far better, he advised Brower, to "stick to your birdwatching" and oppose the Echo Park dam because it would flood a national monument.[11] On the other hand, as historian Mark Harvey points out, Brower was no doubt aware that both Chapman and McKay had already rejected the argument that damming Echo Park would set a precedent for compromising the park system.[12] Brower talked with his father and brother, both engineers, and with Walter Huber, former president of both the Sierra Club and the American Society of Civil Engineers. By the time he had finished perusing and marking up the inches-thick CRSP report, Brower believed he had uncovered not one but two errors in the bureau's presumably impeccable calculations. The conservationists had found their David; he had dug up a stone and was prepared to hurl it.

By the time of the hearing, the bureau had distilled its plans for dams in Echo Park and Glen Canyon down to two options. The first was a relatively lower dam at Glen Canyon, together with dams at Echo Park and Split Mountain, both inside Dinosaur National Monument. This was the three-dam, "low Glen" option. The other was to build the highest dam possible in Glen Canyon and scrub the two in Dinosaur, the "high Glen" option. Brower pointed out that in calculating the difference in evaporation between the two options, the bureau had indeed made a mistake.[13] According to the agency, due to the proposed reservoir's large surface area, the single-dam, high Glen option would evaporate 691,000 acre-feet annually. The smaller, low Glen reservoir would evaporate 526,000 acre-feet, while the two reservoirs in Dinosaur together would send up 95,000 acre-feet, bringing the total evaporated by the low Glen option to 621,000 acre-feet. The difference between the high and low Glen options obviously came to 70,000 acre-feet, but Tudor had said the difference was 165,000 acre-feet.

Somehow the undersecretary had kept in the high Glen model the evaporation from the two Dinosaur dams, even though in that alternative they would not exist. Emboldened by his discovery, Brower went on to point out a second mistake. Applying grade-school arithmetic proportions to the bureau's data, he showed that high Glen would evaporate not 691,000 acre-feet, but 640,000.

Taking all the bureau's mistakes into account, Brower had reduced the difference between the low Glen and high Glen options to 19,000 acre-feet, an amount evidently less than the precision of the bureau's calculations. Armed only with grade-school arithmetic, Brower had turned into a one-man conservation agency, salvaging 146,000 acre-feet of water (95,000 + 51,000) and showing that the difference between the two options was insignificant. The Bureau of Reclamation, Brower told the subcommittee, did not know how to "add, subtract, multiply, or divide." Wayne Aspinall could not believe his ears: "And you are a layman and you are making that charge against the engineers of the Bureau of Reclamation?" Brower responded, "I am a man who has gone through the ninth grade and learned his arithmetic. I do not know engineering. I have only taken Mr. Tudor's own figures which he used and calculated an error to justify invading Dinosaur National Monument."[14]

On the second morning of the hearing, the committee called bureau engineer C. B. Jacobsen to set the bird-watcher straight. "After you have taken your ninth grade arithmetic," Jacobsen advised, "you have to start out with a course in algebra, plane geometry, solid geometry, trigonometry, spherical trigonometry, college algebra, and calculus." Tudor had made a "slight error," yes, but so had Brower, for "evaporation is not a direct relationship of the maximum area of a reservoir." The Sierra Club spokesman had neglected to use the scales on the "old slide stick" that deal with cube power and square root. He had failed to compare "the centroids of the respective volumes," high-falutin' mathematical terms that must have been music to the ears of Aspinall and his western cronies.[15] In an interview decades later, Brower recalled how the "Jacobsen snow job" had reduced the impact of his testimony.[16]

In March 1954, Tudor corrected one of the two mistakes that Brower had pointed out, but sticking to his estimate of evaporation from high Glen, reported to the House subcommittee that the difference in evaporation between the two alternatives was 70,000 acre-feet. But when he appeared before the Senate subcommittee on 28 June 1954, the difference had shrunk again. Tudor's presentation repeated almost word for word the one he had originally given to the House subcommittee, except that the difference between the high and low Glen alternates had now shrunk to 25,000 acre-feet, not far from Brower's 19,000.[17] That prompted Brower to ask, "How much lower can they go—and still be wrong?" The *Salt Lake Tribune,* normally a staunch supporter of the bureau and the proposed Echo Park dam, awarded the bureau's regional engineer its Rubber Slide-Rule Award for "stretching the truth."[18]

In the lengthy questioning that peppered Brower's testimony to the Senate subcommittee, Utah senator Arthur Watkins sought to expose the conservationist as a stranger from a strange land. "You are a public relations man, are you not?" When Brower responded, "No," the senator moved on to a question whose answer he knew would surely be damning: "What is your home address?" The confession came simply, "Berkeley, California."[19]

The eloquent, quick-witted Brower had embarrassed individual bureau officials and scored rhetorical points in committee. But he was an ant biting an elephant. The bureau merely corrected its errors and lumbered on. The upper basin plan gained an invaluable ally when at a press conference on 1 March 1954, President Dwight Eisenhower announced his support, describing the Echo Park and Glen Canyon dams as "key units."[20] Since evaporation no longer favored the bureau's preferred low Glen option, it came up with two new objections to the high Glen dam. First, as a Glen Canyon reservoir filled, water would rise and invade another sanctified area, Rainbow Bridge National Monument, which lay upstream of the dam in a tributary canyon. Second, the porous Navajo sandstone of Glen Canyon would not safely support a more lofty dam. Explained the new reclamation commissioner, W. A. Dexheimer: "Our proposed dam [at] 580 feet high is the maximum (700 above the foundation) that can be built on that site geologically. Evaporation has nothing to do with it."[21]

Why was the bureau so determined to retain the Echo Park dam? Its reservoir would hold only 6 MAF—why didn't the bureau simply jettison the controversial dam? The reason had little to do with geology, evaporation, electrical power, irrigation, river regulation, or flood control. The bureau wanted Echo Park because of another kind of simple arithmetic: the addition and subtraction of votes. Even if Brower did no permanent damage to the project's chances, there were enough potential negative votes to worry Aspinall and his allies. Following the lead of the Republican platform, conservatives regarded the CRSP as creeping socialism. Fiscal watchdogs could tell from back-of-the-envelope calculations that the bureau's vague and preliminary benefit-cost estimates were not worth the paper they were written on. But as always, the powerful California bloc represented the most serious threat. The representatives of the Golden State had been against the project from the get-go, as they would be against any attempt to store water upstream. The blatant tactics of the Californians kept Wayne Aspinall in a state of high dudgeon. He deplored the "vanguard operating for the well oiled and carefully prepared propaganda barrage of the lower basin," going on to complain that "the hungry horde in the lower basin wants every possible drop."[22] Aspinall, a consummate vote-counter, knew that passage of the CRSP was in jeopardy.

With so many likely nays, Aspinall could afford to lose none of the ayes, especially those from representatives of the four upper basin states. The Utah delegation was key. Geography had dictated that the Green River flowed for part of its course through desolate eastern Utah, where almost no one lived and where the state's allocation of Colorado River water could do little good. Utah had long wanted to take water from the Green and Colorado west to the potentially fertile valleys in the midsection of the state and on to the Salt Lake valley. To guarantee that plan, Utah needed its own storage reservoir. Echo Park was ideally situated.

On 9 June 1954, the House Committee on Interior and Insular Affairs voted by the narrowest of margins to send the CRSP, containing both the Echo Park and Glen Canyon dams, on to the full House. Far from being encouraged by the close vote, Brower was incensed. He wrote to Sherman Adams, Eisenhower's assistant, asking the administration to rescind its support for the Echo Park dam. "Mundane men," Brower admonished, "will not know how to take and enforce that stand. It will not please little men. It requires grasp, imagination, vision, and statesmanship."[23] Since the president of the United States had just announced that both dams pleased *him,* the letter surely did more harm than good.

The Interior committee passed the bill that June by a wide margin, but the full Senate did not get around to voting on it during the session. Leaders of both bodies, noting the closeness of the vote in the House committee, decided to let the bill die and resurrect it in the next Congress. That gave the opponents time to change their strategy from the losing game of evaporation to the sanctity of the National Park System, as Leopold had originally advised.

The conservationists were learning to use new weapons. The Sierra Club devoted its February 1954 bulletin to Echo Park. It produced a film comparing the beautiful canyons of Dinosaur National Monument to the now uglified Hetch Hetchy valley. The most persuasive piece was a handsome collection of essays and photographs entitled *This is Dinosaur: Echo Park Country and Its Magic Rivers,* which Wallace Stegner edited with master nature photographer Philip Hyde contributing the illustrations. The book was a work of art and by January 1955, a copy had been given to every member of Congress. Brower's accompanying brochure contained a photo of the ugly mud flats at Lake Mead and a headlined quote from Secretary of the Interior McKay: "What We Have Done at Lake Mead Is What We Have in Mind for Dinosaur."[24]

In the new session of Congress that convened in January 1955, members of both chambers submitted bills with the Echo Park dam included. Aspinall, now chairman of the House subcommittee, found himself with his hands full. The emboldened California members, who had spent hundreds of thousands of dollars on lobbyists, now attacked the basic economics of the CRSP. For the first time in history, the representatives of the Golden State objected to having easterners subsidize westerners. Physician David Bradley had floated the Green and taken the photographs that Brower used in his testimony. Bradley said what many had been thinking, that unless Congress removed the Echo Park dam, conservationists might oppose the entire CRSP. That supporters of the project had to take the threat seriously was a mark of the newfound strength of the conservationists.

Meanwhile, the president of the Sierra Club board wanted to make sure no one doubted its motives. He warned that "the Sierra Club and other conservation groups in California have to be extremely careful not to give the slightest impression that we are fighting the Upper Colorado Project in order to get more water for California. . . . It is not wise for the Sierra Club to get into the argument over the basic economics of the . . . project as a whole."[25] Brower paid little attention, meeting several times with the ubiquitous Northcutt Ely, now reincarnated as an attorney representing Southern California water interests. Brower said he would meet with anyone "who could help us save Dinosaur."[26] The criticism of the faulty economics of the project grew more shrill, accentuated by books and articles. Former Wyoming governor Leslie A. Miller, in an article in *Reader's Digest* titled "Dollars into Dust," wrote that, including interest, an acre of CRSP land would cost at least $2,900, yet be worth no more than $150. He added that, including interest, to irrigate all of the CRSP proposed acreage would cost "over 80 percent of the total present value of all farm lands and all farm buildings in the states of Colorado, Utah, Wyoming, and New Mexico."[27]

The project inspired the *New York Times* to dream of tropical fruit growing on a bare Rocky Mountain top:

> With hardly a ripple, the Senate has coolly authorized a series of public power and reclamation projects for the Upper Colorado Basin states at a cost that even the notoriously inadequate official estimates place at more than a billion and a half dollars. The power will be produced at fantastically high rates; the irrigation per acre will cost many times the conceivable value of the land; and as an incidental by-product the National Park System . . . will be threatened with ultimate destruction. It would of course be possible to grow bananas on top of the Rockies if one wanted to spend the money; but the question is, couldn't this money be spent more advantageously elsewhere?[28]

Consummate vote-counter Aspinall began to figure that keeping the Echo Park dam might just cost more votes than it gained. In June 1955, he announced that the dam would "have to be sacrificed on this altar of opposition."[29] But the Southern Californians, ever perfidious, would not agree. They hoped that with the Echo Park dam included, the entire project would collapse. Aspinall's committee eventually agreed to remove the dam, but the Senate had already approved a version including it. A deal would have to be struck in conference.

On 1 November 1955, forty pro-reclamation governors and House members met in Denver to strategize. The day before, conservationists had placed an open letter in the *Denver Post* threatening to scuttle the entire CRSP unless Congress removed the Echo Park dam.[30] Five years earlier, the threat would have been risible. Now, the conservationists had gathered so much influence that the assembled politicians capitulated. Here was a moment when the conservationists could have allied themselves with the Californians and other opponents and likely defeated the entire project. But they were too honorable. Having made a bargain, the conservationists were bound to keep it. They wrote Aspinall saying that if the bill omitted the Echo Park dam, they would not oppose. Within a few weeks, the secretary of the Interior announced that the bureau had dropped its plan to dam Echo Park.

When Lake Powell filled behind the high Glen dam, a narrow tongue of lake water would protrude up a side canyon, then up one of its tributary canyons, both embedded in a maze of strange, convoluted slickrock, and reach one of the wonders of the modern world, Rainbow Bridge, enclosed in its own tiny national monument. The conservationists had

Figure 12.   First detonation at Glen Canyon, looking downstream, 15 October 1956 (Northern Arizona University, Cline Library, Special Collections and Archives Department)

not merely acquiesced to the high Glen option—they had vigorously endorsed it, the higher the better. They had succeeded in barring the inundation of one national monument only to ensure that a reservoir would invade another.

To prevent that outcome, a month after the politicians accepted the Denver deal, the conservationists came up with an additional demand: the act must also protect Rainbow Bridge. After negotiating with Aspinall, they succeeded in having two key provisions included. Section 1 affirmed that the secretary of the Interior, "as part of the construction, operation, and maintenance of the Glen Canyon Unit . . . shall take adequate protective measures to preclude impairment of the Rainbow Bridge National Monument." Section 3 stated, "It is the intention of Congress that no dam or reservoir constructed under the authorization of this Act shall be within any national park or monument."[31] The first stipulation appeared to save Rainbow Bridge National Monument itself; the second, though it applied only to the CRSP, might supply a precedent that would bar invasion of other parks and monuments.

Figure 13.　High scalers at Glen Canyon, May 1959 (Courtesy of Timothy L. Parks, *Images of America: Glen Canyon Dam* [Arcadia, 2004] and Bureau of Reclamation)

As passage of the CRSP—without the Echo Park dam—drew closer, the leading member of Aspinall's lower basin "hungry horde" made a last-ditch effort to scuttle the project. One matter that could trump any argument for a dam and any benefit-cost analysis was dam safety. The Bureau

Figure 14. Glen Canyon Dam halfway to completion, 4 March 1963 (Courtesy of Timothy L. Parks, *Images of America: Glen Canyon Dam* [Arcadia, 2004] and Bureau of Reclamation)

of Reclamation had first said the Glen Canyon site was unsafe for a high dam, then changed its mind. Here perhaps was an opening. On the morning of 29 January 1956, on the desk of every member of Congress sat a parcel containing a small lump of grayish rock. Accompanying the specimen were instructions to place it in a container with water and wait to see

Figure 15. Glen Canyon Dam complete, 22 November 1963 (Courtesy of
Timothy L. Parks, *Images of America: Glen Canyon Dam* [Arcadia, 2004] and
Bureau of Reclamation)

how long before the rock turned into mud. The offering had come from
Representative Craig Hosmer of Long Beach, who in the company of two
geologists had made a top-secret helicopter trip to Glen Canyon to collect
samples of Chinle shale, a well-known Colorado Plateau rock and one that
would line much of Lake Powell. Hosmer himself demonstrated for
reporters how long it took a piece of the shale to return to the mud from
whence it came: three minutes. The exposure of long stretches of the shale
to the waters of Lake Powell, Hosmer warned, would lead to the same
result but on a vastly larger scale. "The nation's taxpayers," warned the
Californian, would be left "with a billion dollar mud puddle."[32]

In a hearing several weeks later, Hosmer had a fellow House member drop a chunk of the shale into a glass of water and shake it. Again the shale promptly disintegrated. But novice congressman Stewart Udall of Arizona had his own demonstration ready.[33] Udall dropped a drill core of Navajo sandstone, which would serve as the actual footing of the dam, into a glass of water. At the end of his speech, he quaffed the still crystal-clear liquid. One amateur geologist's exhibition having trumped the other's, Udall's amused colleagues quickly passed the bill.[34]

The Echo Park dam was out of the CRSP; the high Glen Canyon option and protection for Rainbow Bridge National Monument were in. Another dam would inundate the vermilion walls of the Flaming Gorge of the Green River near the Utah-Wyoming border. Navajo Dam would rise on the San Juan River in New Mexico near the Colorado border. On the Gunnison River in west-central Colorado, three smaller dams—Blue Mesa, Morrow Point, and Crystal—would comprise the Aspinall Unit. The House-Senate conference worked out the remaining differences and both bodies approved the bill on 28 March 1956. Two weeks later, Eisenhower made it law.

The conservationists of the 1950s had thwarted a dam in Echo Park, in the process inventing techniques that could be used in future battles. But to save one small canyon had taken six years, an enormous effort, and the calling in of many favors. To mount an Echo Park-like battle to defeat each dam the bureau wanted to build in a protected area would be impossible. Wilderness areas without park or monument status had no safeguards and would be sacrificed like Glen Canyon, an offering that conservationists were fast coming to regret. Instead of a strenuous but piecemeal effort, they needed to secure blanket protection for existing wilderness areas and find a way to set aside new ones. That led Howard Zahniser back to an idea he and the Wilderness Society had begun to promote in 1947, to establish a program of national wilderness preserves. "Let us try to be done with a wilderness preservation program made up of a sequence of overlapping emergencies, threats, and defense campaigns," Zahniser urged in an address.[35] In 1964, Congress responded by passing the Wilderness Act, which today protects more than 105 million acres in some six hundred wild areas across the nation.

———

The Echo Park debate was so seminal to the American environmental movement that it is worth a look back with a half-century of hindsight.

The bureau had claimed that evaporation was the critical issue and yet had twice miscalculated how much water the low Glen option would waste. Having uncovered the errors, Brower could hardly let them pass. His audacity in challenging the mighty bureau led to hours of questions and counterchallenges from the assembled politicians and bureaucrats. As the Sierra Club director seemed to recognize at the end of his second day of testimony before the House subcommittee, he had wound up with too little time to make the case for wilderness: "Right now the last impression I am leaving with you is one of figures. I did not come here to do arithmetic. I came here to try to advocate the principle."[36]

The conservationists had no choice but to point out the bureau's errors—but what was the best outcome they could have hoped to achieve? Congress was not about to scuttle the entire CRSP because of an easily corrected mistake or two. To err is human—and in his own testimony, Brower himself had made a mistake. But at least he immediately admitted it, volunteering to demote himself to the eighth grade. Bureau engineers obviously were not incompetent: they had built dams that were wonders of the modern world. After Brower demonstrated that the difference in evaporation between the low and high Glen models was too small to matter, the bureau merely corrected the record and switched to other justifications for damming Echo Park. Aspinall and the other vote counters removed the Echo Park dam only when it appeared that the conservationists, arguing from the principle of wilderness protection, might be able to defeat the entire Colorado River Storage Project.

Had they pushed harder and been willing to abandon their bargain, the conservationists might have been able to save Glen Canyon. But it is hard to feel passionate about something you have never seen. Having few weapons and no firsthand experience of Glen Canyon, the conservationists had no choice but to let the bureau define the debate as about volume and surface area—about geometry, not beauty. Both sides treated Glen Canyon as having no intrinsic worth, as nothing more than a space to be filled, a vacancy that would hold this many acre-feet of water and evaporate that many. Many eloquent words were uttered in praise of the wild beauty of Echo Park. Hardly a one was spoken in defense of Glen Canyon. Both sides agreed that the only attribute of Glen Canyon worth mentioning was how deeply the bureau should drown it: the deeper, the better. The conservationists let the agency choose the ground on which the battle would be fought: reclamation, not wilderness, and they lost Glen Canyon. Brower and his colleagues should at least have visited Glen Canyon. Wallace

Stegner, who had, told Brower: "Glen Canyon. That's where the beautiful material is, the beautiful scenery."[37] Writing thirty years later in a new foreword to a reprint of *This Is Dinosaur,* Stegner had come to believe that had the conservationist movement "known its own strength" and rallied, it might have defeated the dam in Glen Canyon.[38] From Dinosaur National Monument to Glen Canyon is a scenic trip of three hundred miles. For the rest of his life, David Brower regretted not having made that journey in time. In a 1979 interview, he admitted that "I [gave] away Glen Canyon without knowing one cotton-picking thing about what was there."[39]

Most of us looking back at our teenage years find numerous, now inexplicable mistakes of omission and commission. Brower and the conservation movement both had to grow up too fast to avoid mistakes of their own. For that, we can forgive them. Without David Brower and his colleagues, there would likely be dams not only in Dinosaur National Monument, but in the Grand Canyon and who knows what other supposedly protected areas.

ELEVEN

To Have a Deep Blue Lake

"PRECLUDE IMPAIRMENT" of Rainbow Bridge, the bill approving the Colorado River Storage Project stated unequivocally. But almost no one on either side of the Echo Park debate had seen the inaccessible arch. The director of the National Park Service had never been there; neither had David Brower or the Sierra Club board. No one had any idea of what it would take to protect the natural monument from the approaching waters of Lake Powell. It turned out to be a much more difficult and costly job than anyone could have imagined.

The task of protecting the bridge fell to Floyd Dominy, who had become reclamation commissioner in May 1959. Dominy called Lem Wiley, head of construction at Glen Canyon, to say, "Lem, you and I got to go in there and decide what we're gonna do to live with this law."[1] The two westerners rode horses twenty miles through some of the most corrugated and impenetrable terrain on the continent. Not many decades earlier, merely getting to Rainbow Bridge was impossible without a guide, and there was only one. Dominy and Wiley quickly saw that the only way to protect the bridge was to sandwich it between two dams, like the pairs of cofferdams that had protected the Hoover and Glen Canyon dam sites. Topography dictated that the downstream dam at Rainbow Bridge would have to be at least 150 feet high and 100 feet wide. Unfortunately, a dam below the monument would not only prevent water from rising upstream.

Figure 16. Floyd Dominy, archetypal commissioner of reclamation (Courtesy of Bureau of Reclamation)

It would also impound water flowing downstream and back it up under the monument, creating the very danger the lower dam had been designed to prevent. To avoid that outcome, the bureau would have to build a second, fifty-foot-high dam upstream of the bridge, sandwiching the arch between the two.

But the upper dam would also block water running down the creek toward the arch and therefore would require tunnels to divert that water into an adjacent canyon. During the rainy season, water would still collect between the two dams. To remove it, the bureau would have to install large diesel pumps. Two dams, diversion tunnels, heavy equipment, pumps, facilities for workers all in some of the most fragile country in the West, which had neither roads nor power. When the estimated cost of protecting Rainbow Bridge reached $15 million, the director of the National Park

Service moaned, "What an unholy mess they are going to make of a once wonderful national monument."[2]

After a ride that left him so sore he later joked he had "walked out and carried the horse," Dominy had his staff build a topographic model of Rainbow Bridge.[3] It confirmed that protecting the arch would necessitate a major construction project, one that would likely do more harm than would a narrow tongue of lake water protruding into the monument. Dominy had asked Aspinall to have the CRSP legislation amended to remove protection for Rainbow Bridge, but the shrewd Coloradoan had declined, fearing that if he re-opened the act, opponents of the project would have a second chance to defeat it and this time might succeed.[4]

Dominy remembered that although "according to the law . . . we had to ask for money to build these structures," he had "advised the appropriation committee that they would be foolish to give it to me." Dominy had told the senators, "If you'll refuse to give me the money, then we can't build the structures and we'll let the water back into that monument."[5] In 1960, 1961, and 1962, Congress approved Department of the Interior budgets only after the addition of language specifically denying funding to protect Rainbow Bridge National Monument.[6] Washington, D.C. has more than one way to skin a cat.

Congressman Stewart Udall of Arizona decided he ought to see the bridge for himself. He already knew the country around it. His maternal great-grandfather was John D. Lee, Mormon fugitive and operator of the first ferry across the Colorado River at the site named for him. John D.'s son—Udall's grandfather—had led a small Mormon band to tiny St. John's in eastern Arizona. In the election year of 1960, Udall, accompanied by his two sons and Congressman John Saylor of Pennsylvania, launched rafts onto the Colorado at Hite, the old Glen Canyon gold-mining settlement, and set off downstream. With Glen Canyon Dam still under construction, the waters of Lake Powell had not yet risen high enough to prevent them from floating the same quiet waters that John Wesley Powell had described nearly ninety years before. After a leisurely, rapid-free trip, the party reached the mouth of Forbidding Canyon and hiked up it six miles to its small tributary, Bridge Canyon, home to Rainbow Bridge—the sacred Nonnezoshi of the Navajo. The visitors climbed atop the arch, walked beneath it, and hiked a mile up Bridge Canyon to see for themselves how far the waters of Lake Powell might invade.

When they returned, Udall reported to Aspinall that Rainbow Bridge "was unquestionably the most awe-inspiring work of natural sculpture anywhere in the United States."[7] He set forth the dilemma:

The one overriding principle of the conservation movement is that no works of man (save the bare minimum) should intrude into the wonder places of the National Park System. A corollary of this principle is that even the waters of a man-made lake or reservoir constitute an unwarranted Park invasion. Building either of the two proposed dams near the artificial "boundaries" of the Monument would sacrifice the cardinal princip[le] in order to save its corollary.[8]

The simplest way to prevent the waters of Lake Powell from entering Rainbow Bridge National Monument would have been to maintain the lake level below the mouth of Bridge Canyon. But to build a cash-register dam and then keep its reservoir less than full ran counter to the grain of federal reclamation and power generation. After all, the bureau had planned to inundate Echo Park entirely, not just allow a sliver of water to extend barely into it. To allow conservationists rather than power managers to determine the level of one of the bureau's largest reservoirs would have set a dangerous precedent. Moreover, the restriction would limit the elevation of Lake Powell to about 3,600 feet, roughly corresponding to half capacity. Not only would that reduce the head on the generators, it would leave little cushion for drought protection beyond the volume of water that Glen Canyon Dam was required to release each year.[9] And power generation was the true purpose of Glen Canyon Dam.

At the time he wrote to Aspinall, Stewart Udall was a junior congressman whose opinion counted, if at all, only because he was a native of the Colorado Plateau. A few months later, everything had changed. President John F. Kennedy appointed Udall, who had campaigned hard for him, as secretary of the Interior. That disappointed Aspinall, a fellow Democrat who thought he had earned the right to a promotion. Udall was no longer another, necessarily deferential, member of Aspinall's committee: he was the very person whom legislation directed to "take adequate protective measures" on behalf of Rainbow Bridge National Monument. Udall was bound to disappoint one side or the other—and likely both.

The harder it became to figure out how to protect Rainbow Bridge and the longer it took, the higher David Brower's dander rose. In a 1960 article, he warned Sierra Club members, "Something that dollars cannot replace—the national park system—is being endangered again to avoid making good on a promise accepted in good faith." He urged Udall to "direct that further construction at Glen Canyon Dam cease until funds

Figure 17. "Mr. Chairman" Wayne Aspinall points with pleasure as the gauge shows Lake Powell reaching full pool. (Bureau of Reclamation)

are provided to allow him to live up to the agreement made in good faith by the administration and the Congress."[10]

If the pledge to protect Rainbow Bridge could be set aside so easily, so could the agreement not to invade Echo Park—and who knows what other promises. Udall's appointment had pleased Brower at first, as many worse choices, including Aspinall, were available. But as the Interior secretary and Congress continued to dither, Brower's patience wore thin. So did the patience of the Sierra Club board, which tried to have Brower fired. When the effort failed, the board instructed Brower that "no statement should be used that expressly, impliedly, or by reasonable inference criticizes the motives, integrity, or competence of an official or bureau." Then came an admonition that was anathema to the uncompromising Brower: "In publications, objectivity can best be achieved by presentation of both sides of a controversy." Brower ignored what he called the "gag rule."[11]

In March 1962, with Glen Canyon Dam rising higher and the day when the gates to its diversion tunnels would slam shut now only months away, in an open letter in the *Sierra Club Bulletin* and in national newspapers, David Brower made his attack personal:

> *Preclude* impairment, the law says. It doesn't say to plead excessive cost. Or to hustle through some kind of "geological whitewash." Or to arrange series of show-me trips to lead editors and congressmen into believing that protection is just too much load on taxpayers and would tear up the countryside with roads and scars. . . . And when the law says *preclude impairment,* it spells out in unmistakable words: "no dam or reservoir . . . shall be within any national park or monument." Not maybe. Not yes but. Just NO. . . . If Rainbow is not protected, it is not your subordinates who will be responsible. It is you. You, Secretary Stewart L. Udall, the man who dared to have a dream that others hadn't the courage or boldness to dream. And President John F. Kennedy, who you let think your dream was worth dreaming. Don't let him down. Don't let yourself down. Nor us.[12]

Udall declined to respond in kind, pointing out that not only had Congress denied his budget requests for the protection of Rainbow Bridge, it had added riders to the annual appropriations acts that stated that "no part of the funds appropriated for the [CRSP] shall be available for construction or operation of facilities to prevent waters of Lake Powell from entering any national monument."[13]

Since the legislators refused to protect the bridge, conservationists had only one remaining option: litigation. But did they have legal standing to sue the United States government? To find out, on 12 December 1962, the National Parks Association sued in the U.S. District Court in Washington, D.C. for a preliminary injunction to prevent Udall from closing the gates at Glen Canyon.[14] A possible legal precedent was a Massachusetts ruling that a private citizen did have standing to compel a government official to carry out the duties of that official's office. But the district court judge failed to accept the precedent and both the federal Circuit Court of Appeals and the Supreme Court upheld. Without further legal recourse, on 19 January 1963 Brower telegrammed Udall urging him to keep the gates at Glen Canyon open. Instead, the secretary told Dominy to close the gates on the west diversion tunnel two days later, as scheduled. By this time, the river had risen thirty-four feet behind the huge dam. Only a last-minute stay of execution could save Rainbow Bridge.

As the engineers at Glen Canyon prepared to seal the diversion tunnel, Brower boarded a plane for Washington. Once there, he planted himself outside Udall's office, hoping that a final personal appeal would persuade the Arizonan to grant Rainbow Bridge a reprieve. But the secretary was otherwise engaged. Udall and Dominy were on their way to a press conference, where the pair had something important to announce. Udall invited Brower along and as he walked the hall, the conservationist's heart must have sunk. He knew that Udall and Dominy were not going to announce a plan to save Rainbow Bridge. The pair had something new in mind and whatever it was, conservation was unlikely to benefit.

When he heard their plan, Brower was aghast. With Glen Canyon on the way to oblivion and water rising toward Rainbow Bridge, Udall and Dominy had already moved on to their next target: damming the Grand Canyon—twice. One dam would go up in Marble Canyon and the other downstream at Bridge Canyon (this Bridge Canyon has the same name as the small one under Rainbow Bridge). Decades before, LaRue had proposed dams at those very sites; in 1949 Carl Hayden had tried to get the dams funded and the Sierra Club had voted its approval. At age eighty-eight, Dominy had not forgotten: "The Sierra Club had agreed that the Bridge Canyon dam should be built. They're on record in writing. They also had agreed before the Dinosaur National Monument was expanded that they had no objection to a dam in that area of the river. So they're not very pure on the history of their support for these projects."[15]

By the time of the Udall-Dominy press conference, Brower had learned the price of compromise. To save Echo Park, he and the other conservationists had sacrificed Glen Canyon. Now, in spite of the wording of the CRSP legislation, water was about to invade Rainbow Bridge. Listening to Udall and Dominy, Brower must have realized that the Echo Park victory had established no principles and set no precedents. Far from dropping any of its plans to dam national parks and monuments, Dominy's agency, supported by a true son of the West, Stewart Udall from St. Johns, Arizona, was expanding them. Do a deal to save Echo Park—and lose Glen Canyon, Rainbow Bridge, and, if the Dominys of the world had their way—the Grand Canyon! What next? Yellowstone and Yosemite? Why not—both were on the bureau's list. Brower made up his mind that he had compromised his last, a vow he kept until his death in 2000.

Brower and the Sierra Club now came up with a new strategy. In the summer of 1963, the club published *The Place No One Knew: Glen Canyon on the Colorado*. Eliot Porter, whose color photographs adorned the book, proved as adept with words as images: "It is reflection that imparts magic to the waters of the Glen Canyon and its tributaries. Every pool and rill, every sheet of flowing water, every wet rock and seep—these mirror with enameled luster the world about. . . ."[16]

Brower made sure Udall got one of the first copies of *The Place No One Knew*. In the third sentence of the foreword, the Sierra Club director regretted that "the man who theoretically had the power to save this place did not find a way to pick up a telephone and give the necessary order."[17] Udall could well have retorted that when it would have made a difference, the conservationists had not found time to visit either Glen Canyon or Rainbow Bridge. Blame was in ample supply.

In 1964, the club published *Time and the River Flowing*, recording in words and the beautiful photographs of Philip Hyde one of the last raft trips through the Grand Canyon before it fell under the control of the power managers at Glen Canyon Dam. Recognizing a good idea when he saw it, rough and tumble Floyd Dominy revealed a hitherto hidden literary side by writing, and illustrating with his own photographs, *Lake Powell: Jewel of the Colorado*.[18] The frontline of the battle for wilderness had reached the coffee table. David Wegner points out that Dominy began his career as an agricultural agent persuading reluctant, independent Wyoming farmers to sign on to government water projects.[19] The commissioner had learned how to frame a convincing argument.

A copy of the bureau's *Jewel* cost only seventy-five cents, but plenty were available for free. Dominy distributed the book to members of Congress and their staffs, to newspapers and magazines, and to anyone likely to be friendly to reclamation. "God created both man and nature. And man serves God," Dominy wrote. "But nature serves man. . . . To have a deep blue lake, where no lake was before, seems to bring man a little closer to God."

Illustrating the irony of the struggle for the Colorado River, even Lake Powell's strongest opponents had to admit that the azure lake, set against the buff, tan, and ocher sandstone bluffs, was irresistibly beautiful. In *The Place No One Knew*, Brower admitted that Lake Powell, "before it fills and drops, will probably be the most beautiful reservoir in the world, even though the best has gone under."[20] Stewart Udall, in the foreword to *Lake Powell: Jewel of the Colorado*, opined that "once in a blue moon we come upon almost unbelievable beauty. Such was my reaction at my first sight of Lake Powell . . . an exciting new concept of conservation: Creation of new beauty to amplify the beauty which is our heritage."[21] Ardent conservationist Wallace Stegner agreed, but only up to a point: "Though they have diminished [Glen Canyon], they haven't utterly ruined it. Though these walls are lower and tamer than they used to be, and though the whole sensation is a little like looking at a picture of Miss America that doesn't show her legs, Lake Powell is beautiful. It isn't Glen Canyon, but it is something in itself. The contact of deep blue water and uncompromising stone is bizarre and somehow exciting." But then Stegner reached his limit: "In gaining the lovely and the usable, we have given up the incomparable."[22]

Udall and Dominy could propose, but as they well knew from recent experience at Rainbow Bridge, only the Senate appropriations committee could dispose. Its chairman was the indefatigable Carl Hayden. Born in 1877 in Hayden's Ferry, now Tempe, Arizona, he was already thirty-four years old when Arizona became a state. Hayden began his political career as treasurer and sheriff of Maricopa County and in 1912 became his state's first member of the House of Representatives. As county sheriff, so the story went, armed with only an unloaded pistol, Hayden had faced down a bank robber with a loaded one, an exploit that would still guarantee election in any western state. Hayden served in Congress until January 1969, setting the record for longevity until Senator Strom Thurmond broke it. While he chaired the Senate appropriations committee from 1955 until his retirement, four presidents occupied the White House. During those years, no one had more influence on federal spending than the old sheriff.

Throughout his Methuselah-like career, one subject had preoccupied Hayden: to ensure that his arid state did not go thirsty. All the way back in June of 1922, during the hearings on the Swing-Johnson bill that led to the Colorado River Compact, the lanky young congressman had proposed diverting Colorado River water to central Arizona via a canal below the high dam that he knew would eventually have to rise below the mouth of the Grand Canyon. In 1943, Senator Hayden quietly obtained an appropriation to allow the Bureau of Reclamation to inventory water projects in the lower basin. The linchpin would be a hydropower and storage dam at Bridge Canyon. Later the bureau added the Marble Canyon dam to the scheme. By the mid-1940s, with no way to take its share of Colorado River water, Arizona had already begun to deplete its groundwater and its reservoirs. Something had to be done, even if it did mean damming the Grand Canyon twice at a cost of $845 million, or $7 billion in today's dollars.[23]

Arizona needed power dams in the Grand Canyon because the Colorado River below Hoover Dam lies at a substantially lower elevation than the fields and cities of central Arizona. What could lift water, a heavy liquid indeed, thousands of vertical feet? To a bureau engineer, the answer was obvious: the tremendous fall of water through the Grand Canyon. Where the Colorado River enters Marble Canyon, it stands at about 3,500 feet elevation; some three hundred miles downstream it has dropped to only four hundred feet, giving the river an average grade of over ten feet per mile through that stretch. That steep descent would spin the turbines of a hydropower dam so fast that no dam-builder could resist.

Those who might object to damming the Grand Canyon would find that once again, Hayden had anticipated them by decades. The language that he had insisted be added to the Grand Canyon Act of 1919 gave the bureau the authority to dam the Grand Canyon—if and when Congress appropriated the money. The plan that Udall and Dominy announced was similarly clever in that the Bridge Canyon dam would sit below the lower boundary of the national park, while the Marble Canyon dam would lie just above it, making the park the filling in a "dam sandwich," as author Marc Reisner put it.[24]

By the 1960s, most Americans were unaware that conservationists had won a battle over a remote canyon called Echo Park. The Grand Canyon was a different story. Though only 1,500 people had rafted the chasm since John Wesley Powell's first voyage in 1869, in 1963 alone a million and a half visited the rim of the Grand Canyon.[25] In the years since Congress established the national park, a total of over twenty-one million had come to the Grand Canyon and few had forgotten the experience. Just as important, countless other Americans believed that they knew the Grand Canyon from photographs, films, magazine articles, and family slide shows of "our trip out West." It was not that folks had not been to the Grand Canyon—they had not been there *yet*. The notion that federal bureaucrats wanted to flood the Grand Canyon with not one but two cash-register dams bordered on an obscenity. Would anyone want to visit the Lincoln Memorial and find that the reflecting pool had risen and submerged the Great Emancipator up to his nose? Outrageous. The Grand Canyon was deemed part of our national patrimony and those who tampered with it would do so at their own risk.

The conservationists had learned something important from the Echo Park struggle: they had power, including the power of the press. Both *Life Magazine* and *Reader's Digest* ran articles criticizing the idea of damming the Grand Canyon; so did *My Weekly Reader,* a publication aimed at elementary schoolchildren. While an ad about obscure Echo Park would need to locate the site geographically and explain its importance, the Grand Canyon required no introduction. Brower ran ads in the *New York Times* and the *Washington Post* under the headline, "Now only you can save Grand Canyon from being flooded . . . for profit." The ad went on,

"This time it is the Grand Canyon they want to flood, the Grand Canyon."[26]

The ads ran on the morning of 9 June 1966. A few hours later, Arizona congressman Morris Udall, brother of Stewart, was having a drink at the Congressional Hotel with Sheldon Cohen, commissioner of the Internal Revenue Service. "How in the hell can the Sierra Club get away with this?" Mo Udall asked Cohen.[27] The afternoon of the next day, "a nondescript man in a nondescript suit" showed up at the Sierra Club headquarters to deliver a letter from the IRS.[28] The dispatch informed the club that its lobbying on legislation before Congress had jeopardized its tax-exempt status. Moreover, the club would have to inform its donors that their gifts might not be tax deductible, a ruling that would dry up donations and threaten to put any charity-dependent nonprofit out of business. Adding arrogance to aggravation, the IRS set the applicable date of its punishment not from the future time at which it ruled against the Sierra Club, if it did, but from the date the man delivered the letter.

The *Wall Street Journal* and the *New York Times* denounced the IRS's action, the latter noting the service's "extraordinary departure from [its] snail's pace tradition." Mo Udall later admitted that the tactic had been the biggest mistake he made in the battle.[29] It produced a huge outpouring of letters and telegrams, many from writers who had not seen the Grand Canyon but who did not take kindly to the IRS bullying a small nonprofit whose only sin was to try to protect a national treasure against a callous government bureaucracy. Anyone who knew Brower could have predicted that the tactic would backfire. The next ad ran on 25 July 1966, asking, "How can you guarantee these, Mr. Udall, if Grand Canyon is dammed for profit?" then listing the dam sites in national parks that the bureau was known to have in its sights: Big Bend, Glacier, Grand Teton, Kings Canyon, Mammoth Cave, Yellowstone, Yosemite. Even little Arches National Monument near Moab, Utah, was not too small to escape the bureau's appetite. "Should we also flood the Sistine chapel so tourists can get nearer the ceiling?"[30]

The bureau flinched first by proposing to build the Marble Canyon dam and dropping the one at Bridge Canyon. Then in January 1967, it announced just the opposite: the Marble Canyon dam was off the list, replaced by one nobody had heard of, "Hualapai dam." The name derived from the tribe on the southwest rim of the Grand Canyon who more than a century earlier had guided the first Americans down from rim to river at the entry of Diamond Creek. But where and what was Hualapai dam?

Mo Udall presented the revised bureau plan in the House subcommittee, managing to do so without revealing that Hualapai dam would occupy exactly the same site as the now aborted Bridge Canyon. But no matter what one called it, Hualapai dam would back water into Grand Canyon National Monument, just as Bridge Canyon would have done. Udall and the bureau had a way around that difficulty. The president would simply rescind monument status. Of course, that would imply that the Grand Canyon had not been worth protecting in the first place.

But to the Sierra Club, "one bullet in the heart was as deadly as two."[31] The dam-builders could change the name of a dam and abolish a national monument, but still they intended to dam the Grand Canyon. During the House hearings, Morris Udall, in disbelief at the Sierra Club's obstinacy, asked Brower whether there could not be a "low, low, low Bridge Canyon Dam, maybe 100 feet high, is that too much? Is there any point at which you compromise here?" Brower replied, "Mr. Udall, you are not giving us anything that God didn't put there in the first place, and I think that is the thing we are not entitled to compromise." He added, "We have no choice. There have to be groups who will hold for these things that are not replaceable. If we stop doing that, we might as well stop being an organization, and conservation organizations might as well throw in the towel." Udall responded that he knew "the strength and sincerity of your feelings and I respect them."[32] In spite of losing its tax-deductible status and some of its largest donors, membership at the Sierra Club doubled, with dues from the new members making up much of the financial loss.[33]

The Udall brothers were patient gentlemen; Floyd Dominy was often neither. The Sierra Club director's claims about the extent to which a reservoir would harm Grand Canyon National Park, Dominy said, made him a "damned liar."[34] Decades later, he had not changed his opinion, calling Brower "a sanctimonious bastard."[35] Wayne Aspinall shared Dominy's appraisal. When asked at a conference to have his photo taken with Brower, the Coloradoan declined, "No picture of mine is being taken with that liar! He's been filling the newspapers with a bunch of damn lies!"[36] Luna Leopold, Brower's secret ally, was another who Aspinall believed had crossed him. Once when Leopold was on a commercial flight, he found himself within hearing distance of fellow-flyers Aspinall and Dominy, who said, "Oh, yes. There's that son of a bitch Leopold."[37] The pair of dam-builders had ample reason to doubt Leopold's loyalty. First was the name, with its suspect genealogy.

Second, Aspinall and Dominy likely knew that the hydrologist had been slipping Brower unpublished data from Geological Survey studies. Third, as we will see, in 1958 Leopold demonstrated that Glen Canyon Dam and Lake Powell were unnecessary to regulate the Colorado River. In that study, his scientific analysis of how much water the river could deliver led newspapers to the headline "Leopold Takes Two Million Acre-Feet Out of the Colorado River."[38]

In the heat of the fight for the Grand Canyon, an always-confident Dominy made a rare mistake: he left the country. By the mid-1960s, the Bureau of Reclamation was not merely building dams across the West—it was exporting the gospel of big hydropower dams to nations around the world, helping to fuel a frenzy of global dam-building that seems to have no end. In late January 1967, Dominy left on his annual junket of foreign water projects.[39] About the time the commissioner's flight took off, Stewart Udall called in his staff to tell them that he had changed his mind: he was canceling both Grand Canyon dams. At a press conference the next day, Udall told the entire country.

Udall may have been influenced by his friendship with Luna Leopold, whom the secretary had a habit of inviting to lunch. On one such occasion, Leopold remembered that when he entered Udall's office, he had his feet up on the desk and was chewing an apple. Udall said, "Luna, you've been down the Grand Canyon. Is it worth saving?" Leopold responded, "Well, Mr. Secretary, if you want to save it for a bunch of damn tourists who are going to mess the place up, it's not worth it. But if you are going to make a real name for yourself as a conservationist, which I know you'd like to do, the one thing that will make you famous for the rest of time is if you come out against the dam in the Grand Canyon."[40] And to his everlasting credit, that is just what Stewart Udall did.

With the hydropower dams out of the project, the energy to lift water from the lower Colorado River to central Arizona would come from a huge coal-fired plant on the Navajo Reservation near Page, Arizona, northeast of the Grand Canyon. Coal from near-surface deposits on Black Mesa would power the plant. Drawing its cooling water from Lake Powell, the Navajo Generating Station would produce 2,250 megawatts, all of it running in wires high over Navajo hogans, half of them without electricity.

Before he would agree to support the legislation, Wayne Aspinall insisted that five pet reclamation projects in western Colorado be added. Mo Udall refused to drop the Hualapai dam proposal. But brother Stewart knew how to play politics. He met with key western senators, especially the crucial Hayden, who agreed to remove the Grand Canyon dams as the only way any water was going to get from the Colorado River to the cities of central Arizona. Most importantly, President Lyndon Johnson, who had at first endorsed the dams, announced a change of heart, now saying that if the Central Arizona Project (part of the larger Colorado River Basin Project, enacted in 1968) "includes dams in the Grand Canyon, I will veto it." Johnson, though unable to extract the country from the Vietnam war, had shown through his endorsement of the Civil Rights Act that he could change his mind. But to get him to do so was not easy, requiring an even more stubborn person to whom he would listen rather than attempt to browbeat. There may have been only one such: Lady Bird Johnson, known best for having ugly billboards removed and wildflowers planted along the nation's interstate highways. During the crisis over the Grand Canyon, Lady Bird's Office of Beautification employed a staff member who had been with the Wilderness Society and who now "worked with Lady Bird constantly." She also worked *on* Lady Bird, and Lady Bird worked on LBJ.[41]

When Dominy returned from his foreign excursion, he learned to his disgust that not only was he not going to get the dams he wanted in the Grand Canyon, this life-long water man was going to have to enter the coal-mining and power-plant businesses. Aspinall continued to resist, sending the decision into 1968. But Mo Udall revealed the state of opinion when he admitted that "no bill providing for a so-called 'Grand Canyon Dam' can pass Congress today."[42] In the summer of 1968 the House and Senate conference committee worked out their differences. The most significant section of the legislation was the Central Arizona Project, long sought by Carl Hayden. The bill specifically disallowed study or construction of dams on the Colorado River between Glen Canyon and Hoover dams. But it did include the five pet projects that Aspinall had demanded, which one bureau official would later call "pure trash."[43] Those five projects, if they would do little good for the State of Colorado, would do no harm to Arizona. More ominously, in order to get the California delegation to vote for the bill, the Arizonans had to agree that their water from the Central Arizona Project would have a junior priority to California's receipt of its 4.4 million acre-feet of Colorado River water. One reason for Arizona's willingness to agree was that the bill promised that the state's junior status would be removed when water supplies from the Colorado were augmented by 2.5 MAF annually— though it did not specify the source of the extra water.[44] Both branches of government passed the bill and on 30 September 1968, President Johnson made it law.

—

As the tumultuous 1960s ended and attitudes toward wilderness began to change, Wayne Aspinall, David Brower, Floyd Dominy, and Carl Hayden each left their posts—defeated, fired, or retired. The Colorado legislature had redrawn the boundaries of Aspinall's once-safe seat on the western slope to include more conservation-minded voters from the eastern side of the Front Range. After conservationists placed him among the "dirty dozen" anti-environmentalists, Aspinall lost the 1972 Democrat primary. Following a by-now well-trodden path, Aspinall became a lobbyist. He also founded the ultraconservative Mountain States Legal Foundation, which spawned two future Interior secretaries: Gale Norton and James Watt.

By 1968, Hayden was ninety-one years old and demonstrably senile. When told that he was likely to lose his re-election to Barry Goldwater, the old sheriff had the presence of mind to retire, ending a career that began in a dusty, sparsely settled western territory that would soon hold countless fountains, blue lakes, and scores of golf courses. Where once crawled the Gila monster and scorpion, kayakers row, a fine mist cools mall shoppers, and housing developments seem to spring up overnight. And no one could take more credit than Carl Hayden.

What of Floyd Dominy? One would love to have been a fly on the wall one day in 1969 when he received an unlikely emissary. The experienced commissioner had served under four presidents in ten years. That two were Democrats and two Republicans gave paranoids on both sides reason to question Dominy's loyalty. Now Dominy had gotten on the wrong side of two of the most suspicious and vindictive men in town: Richard Milhous Nixon and J. Edgar Hoover. Their agent was a callow youth of barely thirty who hailed from the State of Wyoming, where Dominy had started his career. To fire Dominy, Nixon had dispatched James Gaius Watt. Dominy knew that the president and Hoover had the goods on him. Nixon had ordered the FBI to investigate each of the senior federal officials that he had inherited from Lyndon Johnson. Dominy's file was inches thick; like scripture, you could find in it whatever you sought.[45] The normally pugnacious Dominy resigned without a fight.

As Aspinall, Dominy, and Hayden departed, Congress continued to pass legislation to shield the environment. The best dam sites were gone and new laws protected the rest. Indeed, Glen Canyon would be the bureau's last great, twentieth-century dam. The Central Arizona Project was far enough along to ensure it would proceed to completion, though coal rather than falling water would power its pumps. In 1975 Congress redrew the boundaries of Grand Canyon National Park to include all nearby Indian lands, ending the prospects for a future Hualapai dam. None of Dominy's many other plans for western water projects came to fruition.

Even Dominy's strongest opponents would have to admit that he missed no opportunity to fulfill what he saw as his mission. Looking back decades later, Dominy said, "I have no apologies. I was a crusader for the development of water. I was the Messiah. I was the evangelist who went out and argued persuasively for the harness of water for the benefit of people. Here out on the river, you won't hear any weasel words from Floyd Dominy."[46]

In the same year that Watt arrived to fire Dominy, the board of the Sierra Club fired David Brower, who had offended too many board members too many times. Their anger boiled over after Brower paid for another expensive ad in the *New York Times* without clearing it with the board.[47] Signaling its lack of confidence, the board removed Brower's spending authority. He and his shrinking number of supporters next lost a fight over the position the club should take regarding a proposed nuclear power plant at Diablo Canyon near San Luis Obispo, California. Now even Brower's best friends on the board began to desert. At the meeting of the board in June 1969, when only five members supported his continuation and ten opposed, Brower quickly resigned.[48] Undaunted, he went on to found new environmental organizations: Friends of the Earth, League of Conservation Voters, Earth Island Institute, and Living Rivers.

But first, Brower headed to Glen Canyon, where he and two companions boarded a raft for a voyage though the Grand Canyon. One of the trio, author John McPhee, recorded the trip with the man he came to call the archdruid of the environmental movement.[49] The third rafter was a big, cigar-chomping man in a western Stetson, none other than Floyd Dominy. Before starting the raft trip, at Dominy's insistence, the party spent a week on Lake Powell.[50]

McPhee's tale of this voyage of strange campfellows is one of his best. Reading his account of the floating debate today, one cannot resist asking not only who won their riverine contest, but who won the larger battle—the conservationists or the reclamationists? By the time the three made their voyage, Congress had passed a stack of legislation to protect the environment, the first Earth Day was just months away, water ran freely through Echo Park and down the Grand Canyon. On the other hand, Glen Canyon lay drowned under hundreds of feet of water; and far up the Green River, reservoir water submerged the vermilion walls of Flaming Gorge. The water that flowed through Echo Park and the Grand Canyon did so only at the command of the Bureau of Reclamation. Fulfilling the goal of Arthur Powell Davis, John Widtsoe, Floyd Dominy and others, the bureau had brought the wild Colorado under man's control. Canyons across the West awaited dams, when the next long drought arrived.

As for the Grand Canyon, one could not always see it. As Dominy was fond of pointing out, haze from the coal-fired, fume-emitting Navajo Generating Station east of Page—the substitute for power dams in the Grand Canyon—often made it hard to see from one side of the chasm to the other. Would a "low, low, low" power dam deep in the Grand Canyon, as Mo Udall had suggested to David Brower, have been worse? The rafter and the environmentalist would surely answer yes; the day-tripper to the South Rim, having difficulty seeing across the Grand Canyon to the opposite rim, might have a different opinion.

When Lake Powell filled, it did protrude under Rainbow Bridge, but it seems to have done no harm to the ramparts of the arch. Ironically, and illustrating once again the dilemma of conservation versus accessibility, far more people have seen Rainbow Bridge by motorboat and a short hike than would have ever seen it on foot or horseback.

The question of who won the larger battle cannot really be answered. In a fight over the environment, one side does not simply win while the other loses. When the Concrete Pyramid wins, the victory may last far beyond the lifetime of any current actors. A high concrete dam in Echo Park would have stood for thousands of years—no one can say how long. A later generation, deciding it wanted to remove the dam, would find the cost prohibitive. Environmental victories, on the other hand, nearly always turn out to be ephemeral. Echo Park and the Grand Canyon do not have dams, but who is to say that when the twenty-first-century water crisis becomes severe enough, proposals to dam them will not resurface? A mountaintop in Appalachia may not be planed off for a strip mine—yet. A wetlands may not be filled in—yet. The Endangered Species Act of 1973 may not have been rescinded—yet. The best environmentalists can hope for is to avoid losing for the time being. Once they do lose, on a human time scale the damage is likely to be permanent. Bernard DeVoto's commentary from fifty years ago remains apt:

> Even when controversies have been formally settled and projects abandoned apparently for good, the park system—and the public trust—is always under the threat that engineers may revive their discarded plans at any time. [Interior] secretaries change, administrations change and formal commitments by the Bureau of Reclamation change, while the itch of engineers to perform costly miracles goes on forever.[51]

Utah physician Richard Ingebretsen presides over the Glen Canyon Institute, whose mission is to restore "a healthy Colorado River through Glen Canyon." In 1997, Ingebretsen paid a call on eighty-eight-year-old Floyd Dominy, finding him in good health on his Virginia farm with only his dog for company. Ingebretsen ostensibly had come to collect a pair of bookends made from Glen Canyon drill cores that Dominy had promised Brower on their famous raft trip. But Dominy had changed his mind. Instead of sending the bookends to his old nemesis, Dominy asked Ingebretsen to photograph him holding the two cores with *Lake Powell: Jewel of the Colorado* sandwiched between, while the old dam-builder laughed at Brower for the camera. He had forgotten nothing.

At a restaurant dinner that night, the talk naturally turned to Glen Canyon Dam and Lake Powell. When told that the movement to drain the lake was serious, Dominy proved unable to resist a large-scale technical challenge, even if it meant tarnishing his jewel. Brower had proposed draining the lake by drilling out the original diversion tunnels. Dominy warned, "You can't do that. It is 300 feet of reinforced concrete." To Ingebretsen's amazement, Dominy went on to show how it could be done: "There is a better way," said the commissioner. "All you have to do is drill new bypass tunnels around the old ones in the sandstone; then you can put waterproof valves at the bottom of the lake. They can be raised and lowered as you need, to let water out."

As Ingebretsen tells it, "With that he pulled over a cocktail napkin and drew a sketch of Glen Canyon Dam, the old bypass tunnels, the lake, the river, and the new tunnels with the waterproof valves that will be used to drain the reservoir. His hands worked busily as he explained what he was

Figure 18. Dominy's dinner napkin sketch showing how to drain Lake Powell (Courtesy of Rich Ingebretsen, Glen Canyon Institute)

sketching." Dominy explained, "This has never been done before, but I have been thinking about it, and it will work."[52]

Ingebretsen was stunned but recovered sufficiently to say, "Mr. Dominy, no one will believe me when I tell them that you drew this. Would you sign and date it?" "Sure I will," Dominy answered, and signed the napkin.

TWELVE

The Biggest Boondoggle

WHY WAS GLEN CANYON DAM BUILT? What were the justifications at the time of the 1950s debate? How sound were those justifications then and how well have they held up? These questions we now have the background to address.

Start with the economics of the Colorado River Storage Project. Remember that the bureau used two procedures to justify new dams. First, recognizing that Congress would be unlikely to approve new projects that would lose money, the agency adopted benefit-cost analysis, estimating the future monetary benefits of a project and comparing them to the construction and operating costs.[1] A sufficiently positive ratio would justify funding. As we will see, benefit-cost analysis provides many opportunities to fudge both numerator and denominator. The second change was river-basin accounting. As explained in chapter 8, the idea was to take a set of smaller, money-losing projects, lump them together with profits from cash-register dams, and turn the overall benefit-cost ratio positive. But why not just drop the money-losing projects, build hydropower dams, and wind up with even higher benefit-cost ratios? Because power generation was not the mission of the Bureau of Reclamation—it was a means to an end. The bureau needed to reclaim land; otherwise it would compromise its core mission and jeopardize its future.

Illinois senator Paul Douglas, a decorated World War II veteran and former professor of economics at Reed College and at the University of

Chicago, was the most effective exposer of the flawed finances of the CRSP. His testimony before the Senate Subcommittee on Interior and Insular Affairs was a tour de force of logic and a revealing primer on the economics of large reclamation and power developments. Douglas was a liberal Democrat who did not object to projects that benefited only one region of the country, as long as the national taxpayer did not have to foot the bill. But of the residents of the upper basin states, Douglas said, "It is too bad that these fine people live in a semiarid region, with a river running through deep canyons. We are sorry for them," he went on, "but I do not think that creates for them a perpetual claim on the Public Treasury." The Illinois senator would not have objected to the bureau's early irrigation projects, because they were at lower, warmer elevations. But the CRSP would irrigate higher, upstream fields, where "there is a short growing season, costs per acre are enormous, benefits per acre will be comparatively low."[2]

At the time of the CRSP hearings, the government was cutting national wheat production by twenty million acres; the secretary of Agriculture wanted to cut other crops by another twenty million. If, in spite of the evident surpluses, Congress was determined to invest the taxpayer's money by irrigating more land, Douglas pointed out, the fertile Midwest would be a cheaper and more productive place. To make matters worse, 95 percent of the land that the Colorado River Storage Project was to irrigate would grow alfalfa and other grasses to feed livestock, "about the most unprofitable use which could be made of irrigated land."[3]

Had the CRSP been cheap enough, that it benefited mostly alfalfa farmers might not have mattered. But the project was not cheap: it was far more expensive than any the bureau had ever undertaken. In making this point, Douglas matched wits, eloquence, and cogent economic analysis with the best Senate orators and bureau engineers.

Douglas calculated the true cost of the CRSP dams from three perspectives: construction, power generation, and irrigation. According to the plan the bureau submitted to Congress, to build Glen Canyon Dam, for example, would cost $421 million. The bureau charged $370 million of the total against power generation and the rest against irrigation. The dam was to have a generating capacity of 800,000 kilowatts, bringing the construction cost per kilowatt to $463. Using the same reasoning, a kilowatt at Echo Park would cost $640, and in another component of the CRSP, the Central Utah Project, $765. These costs were three times as high as in the Tennessee Valley Authority, four times as high as at Bonneville on the Columbia, and almost five times as high as at Hoover Dam.

Douglas summed up: "We are being asked to spend $656 million [the total allocated for construction in the CRSP] in about the worst place in the United States where hydroelectric power could be developed."[4]

Next Douglas turned to operating costs. Once the dams were up and running, how much would it cost to generate a kilowatt-hour of energy? Engineer Ulysses S. Grant III had shown that at Glen Canyon to generate one kilowatt-hour would cost between 4.2 and 4.7 mills (a mill is one-thousandth of a dollar; these were the costs in the mid-1950s). At Echo Park, the cost would come to 6 mills and on the upstream dams would be higher still. Yet the bureau reported that it would have to sell project power at 6 mills per kilowatt-hour, leaving Glen Canyon the only CRSP dam that would show a net profit. As with construction, the cost to generate CRSP power was much higher than elsewhere. The average for the Tennessee Valley Authority dams was 1.1 mills per kilowatt-hour; for Bonneville, 0.6 mills, and so on.

Finally, Douglas calculated the cost of the dams per acre of land irrigated. Logically, those costs should include forgiven or deferred interest, which is a real expense to someone; in this case, to the taxpayer. Douglas began by reviewing the steadily lengthening period of interest forgiveness on federal reclamation projects. In 1914, the original ten-year interest vacation had been increased to twenty years—after an initial five-year period for development—making a total of twenty-five years. In 1926, Congress raised the period to forty years and in 1939 increased it again, to fifty years. Under the CRSP as proposed, the interest-free period would climb to sixty years and thus far outlive most farmers.

Some of the individual irrigation projects were blatant money-losers. Consider Paonia, Colorado, west of today's ski resorts at Aspen and Crested Butte. Residents of the sparsely settled region had been trying for years to get Congress to give them a water project. With the CRSP ready to launch, it appeared that their thirst was going to be assuaged. The bureau allocated $6.79 million of the total construction cost in the CRSP to Paonia. The water would irrigate 7,153 acres, bringing the cost per acre to $949. But that did not include interest, which according to the bureau, given the sixty-year repayment period, would amount to 125 percent of construction costs. Thus including interest, to irrigate one acre at Paonia would cost the national taxpayer $2,135. Douglas estimated that after irrigation, that acre would be worth no more than $150. Paonia was far from the worst: to irrigate an acre under the Central Utah Project would cost nearly $4,000.

In his peroration, Douglas noted that when reclamation funds are repaid, they do not return to the federal treasury, but fund more reclamation projects. Thus taxpayers never recoup their investment—their dollars fund new projects, whose investment is never repaid, and so on, in a kind of hydraulic Ponzi scheme. Revealing he was a better economist than entertainer, Douglas paraphrased an old miner's song: "O, my darling Clementine, the irrigation money once appropriated is lost and gone forever; no matter how dreadfully sorry we may be, it will avail us nothing."[5] Then came his conclusion:

> The Bureau of Reclamation . . . is trying to dam all the streams, install power, provide irrigation, displace all the gorges, and turn scenery into the growing of hay and corn and the production of electricity—all well in their own way, but not all of life, because men must live by something more than beefsteak and fruit.
>
> The upper Colorado project would cripple the reclamation features of the country and tend to transform the Nation physically, into a placid, tepid place, greatly unlike the wild and stirring America which we love and from which we draw inspiration.[6]

To Douglas, the Colorado River Storage Project was "the biggest boondoggle I have ever heard of."[7]

———

Douglas also had it on good authority that no matter what the bureau's cost estimates showed on their face, the underlying numbers were unreliable. The report of the Committee of Special Advisors on Reclamation, all the way back in the early 1920s, had concluded that "the data and evidence before the committee disclose that in almost every project undertaken by the Reclamation Service the ultimate cost of the project greatly exceeds the estimates." Moreover, "For some projects . . . the ultimate cost was more than the value of the water to the land."[8] As experience accumulated, it became apparent that not only did the agency routinely underestimate costs, it did so by at least a factor of two. One source for that figure was noted reclamationist and former president Herbert Hoover, whom President Harry S. Truman had asked to chair a panel to identify ways to reduce the number of federal agencies and improve their efficiency. The Second Hoover Commission reported in 1955 that though the bureau had estimated the cost of the ninety projects it had under way at $1.6 billion, by the time they were completed, the set of projects cost more than twice that amount.[9] The giant Missouri Basin Project and the Colorado–Big Thompson Project both cost four times their original estimates. Had Congress been better advised, said the commission, it would not have authorized some of the projects.[10] Hoover, who had been observing federal reclamation since his days as chair of the Colorado River Commission, described the work of the bureau and the Army Corps of Engineers as a "great hodgepodge of duplication, overlapping and unbelievably extravagant planning."[11]

The commission went well beyond pointing out the cost overruns of the bureau and the Army Corps of Engineers. So alarmed had this independent body become about the "swollen, pork-fed condition of our two principal potential patients" that it recommended stripping the corps of its nonmilitary functions, shutting down the Bureau of Reclamation, and even replacing the Interior Department with a new Department of Natural Resources that would subsume all dam-building.[12] Even before the commission report appeared, the water establishment began lobbying against these radical proposals; soon it had cast them into oblivion.

Nearly twenty years after the Echo Park debate, the bureau's inability to estimate costs had gotten worse. *Damming the West* was one in a series of reports under the imprimatur of corporate thorn-in-side Ralph Nader. Drawing on the 1969 report of the reclamation commissioner, the book reported that the bureau had estimated that 120 projects, originally slated to cost $1.4 billion, would wind up costing well over $4 billion. The Central Arizona Project, the last big effort the bureau would complete in the twentieth century, was originally slated to cost $832 million. By the time it was finished in the 1990s, the Central Arizona Project had cost $3.6 billion.[13]

Some observers, even some bureau insiders, believed that the consistent cost underestimates were no accident. In a 1949 article in the widely read *Saturday Evening Post,* Hoover commission member and former governor of Wyoming Leslie Miller charged that the bureau and the corps both "apparently deliberately" underestimated the costs of their projects, "bamboozling Congress into easy acquiescence."[14] Six years later, at the height of the Echo Park battle, in an article in *Reader's Digest,* Miller described eight bureau projects, including the California Central Valley and Colorado–Big Thompson projects, that the bureau had estimated would cost a total of just over $1 billion. Instead, the projects had cost over $4 billion and the largest, the Missouri Basin Project, was not yet finished at the time Miller wrote.[15]

The most damning comments came from Daniel Dreyfus, the bureau's "house magician" and water project planner.[16] In 1981, Marc Reisner interviewed Dreyfus, who by then was staff director to the Senate committee on energy. Reisner asked him to reflect on his time at the bureau, particularly on the five western Colorado projects that Wayne Aspinall had insisted be added to the Central Arizona Project. "Those projects were pure trash," Dreyfus admitted. "I knew they were trash and Dominy knew they were trash. The Office of Management and Budget had just bounced Animas–La Plata [one of the five]. I had to fly all the way out to Denver and jerk around the benefit-cost numbers to make the thing look sound."[17]

It is worth a moment to consider Animas–La Plata, for it provides a case study of how hard it is to kill a federal reclamation project, no matter how much it deserves to die. By the 1930s, farmers and irrigators along the La Plata River southwest of Durango, Colorado, found themselves with potentially fertile fields but too little water. Over the continental divide to the east ran the Animas River. Transferring water between basins, especially through the middle of a mountain, was just the sort of technological challenge the bureau sought. But the project was so blatant a money-loser that for years no Colorado politician would support it. Finally the time came when Aspinall judged that he had the power to gain approval of any project. Hayden and his fellow Arizonans had been trying for decades to get the Central Arizona Project authorized, handing Aspinall a lever. Animas–La Plata mattered to a relative handful of people—the Central Arizona Project was to be the life-blood of the state. Unless his five Colorado projects were included, Aspinall insisted, he would block the Central Arizona Project. As described in chapter 11, he got his way.

Animas–La Plata was so complicated and demonstrably unsound that more than three decades went by before the first shovel bit in. As originally conceived, power-gobbling pumps would lift 25 percent of the Animas River 1,000 feet uphill and down into a set of reservoirs that would store the water and send it on through forty-eight miles of canals and pipelines to fields growing alfalfa and other low-value crops on the La Plata. Half the total expenditure would go to overcoming gravity. Then, after costing roughly $5,000 to irrigate, each acre would produce about $300 worth of crops.[18] The bureau's own benefit-cost analysis showed the project would lose money, leading the Interior Department to say that "the Animas-La Plata project is not economically feasible."[19] By 1995, the project still not begun, *U.S. News and World Report* described it as the "last surviving dinosaur from the age of behemoth water schemes."[20] Originally slated to cost $710 million, in 1990 Congress scaled down the project to $338 million.[21] By 2003, the estimate had risen to $500 million. Phil Doe, a former bureau official who now heads the Citizens' Progressive Alliance that opposes Animas–La Plata, said that the real news is that "the Bureau of Reclamation knew the estimates to be bogus back in 1999 when they submitted them to Congress." Doe estimates that Animas–La Plata will cost more than $1 billion.[22]

Animas–La Plata also illustrates the rising power of Indian water rights. Both the Animas and the La Plata flow through the Southern Ute Reservation; the La Plata drains the Ute Mountain Reservation to the west. Because Indians and white settlers both owned land that Animas–La Plata would irrigate, the rights of the tribes had to be taken into account. The Colorado state engineer feared that when the inevitable drought arrived, long-standing court decisions would show that the Indians had prior rights to virtually all Animas–La Plata water. To avoid the legal battle that could then ensue, the tribes, the irrigators, the State of Colorado, and the federal government negotiated a settlement giving most of the project water to the two tribes, which number about 3,000 people. The rights of other tribes, for instance, of the Navajo to Colorado River water, remain to be quantified and adjudicated. If they are judged to be substantial, an already overextended Colorado River will be even further in deficit. Indian water rights are the slumbering Monstro of the Southwest.

Now back to the bureau's cost estimates. The agency routinely underestimated them for at least three reasons. First, doing so raised the benefit-cost ratio, not only justifying a project but making it appear that Congress would be irresponsible not to invest taxpayer dollars at such a favorable rate of return. Second, the lower the apparent cost, the less money needed from tight budgets to get a project off the ground. As Wyoming governor Miller pointed out, "after the first batch of concrete is poured, the Engineers and the Reclamationists always can come back . . . for a supplement appropriation."[23] Third, in government work, cost overruns often carry no penalty. It is next to impossible to find a bureau project that Congress halted for any reason, including higher than expected costs. Once begun, a federal reclamation project takes on a life of its own, achieving an almost supernatural longevity. Even when a federal agency admits that a project has a markedly unfavorable benefit-cost ratio, as in Animas–La Plata, the work goes ahead anyway. The bureau and its "rival in crime," as author Marc Reisner labeled the Army Corps of Engineers, appear to be impervious to control by anyone, including the Congress.[24] In 2007, in spite of the role the corps played in the debacle of Hurricane Katrina and the failure of the New Orleans levees, Congress approved a $23 billion water bill without requiring any significant reform of the corps. Once such an opportunity is past, it may not come again for decades. To sum up: large-scale reclamation projects can only be stopped before they start. Even then, as Animas–La Plata shows, Rip Van Winkle-like, they may slumber for decades before they stir.

The criticisms of Paul Douglas should have helped defeat the CRSP, but they only delayed the inevitable. It came down to votes and the pro-reclamation forces had them. William Weld, former governor of Massachusetts, once explained how such matters work: "We don't need to be logical about this. This is politics."[25] Of Aspinall's five pet projects, two were built, Animas–La Plata is under construction, and the bureau dropped two as "infeasible."

If it is hard to estimate the costs of large reclamation projects within a factor of two, imagine how easy it is to overestimate the benefits, which lie decades in the future. One exception is the benefit known as regulation: the capability of a dam and reservoir to store and release water for irrigation and flood control, smoothing out a river's natural ebb and flow. Common sense tells us that a reservoir that stored, say, one-tenth a river's annual flow would offer too little regulation; one that stored ten times the flow would likely provide more regulation than needed. But evaporation complicates the calculation. The more water a reservoir stores, the larger its surface area and the more water it loses through evaporation. As storage increases, a point is reached at which the loss due to evaporation more than exceeds the gain in regulation, obviating the case for the additional storage.

As the bureau began to erect Glen Canyon Dam in 1958, Luna Leopold analyzed the hydrology of the Colorado River for the court-appointed "special master" in *Arizona v. California*. Leopold concluded that "total reservoir capacity in excess of about 40 MAF would achieve practically no additional water regulation."[26] Since Lake Mead already held 28 MAF of the optimal 40 MAF, Leopold concluded that "if reservoirs with capacity beyond an additional 10 million to 15 million acre-feet are constructed in the upper Colorado River basin, evaporation loss will thereafter offset the hydrologic benefit of the regulation so achieved."[27] Flaming Gorge,

Figure 19. Master hydrologist Luna Leopold (*standing at center, facing camera*) taking measurements on the Colorado River in the Grand Canyon (Courtesy of U.S. Geological Survey)

Navajo, and the other smaller dams in the CRSP would have added the extra storage that Leopold deemed sufficient. In other words, in order to regulate the Colorado River, Lake Powell was unnecessary.

Leopold demonstrated that the river's variable flow made it hard to say with precision how much water it could reliably deliver. The amount might be as high as 17 MAF, but it might be as low as 13 MAF—2 MAF less than the total of 15 MAF allocated to the two basins. As noted earlier, this led western newspapers, focused on the low end of his projection, to accuse the hydrologist of taking two million acre-feet from the Colorado River. It also led the governors of several western states to send delegations of engineers to Washington to ask the Interior secretary, Fred Seaton, to fire Leopold. Seaton responded, "The Geological Survey is a scientific agency in which they're supposed to give their best opinion. That's what Leopold did, and I'm backing him."[28]

In his report, Leopold made another point—a near tautology—that nevertheless bears repeating as proposals for new dams inevitably arise. "It would be physically impossible to add, on the average," wrote the master hydrologist, "more water to storage than the average annual flow of the stream." In other words, *on the average,* no more water than the natural flow of a river can enter a system of dams and reservoirs and *on the average,* no more can be removed without eventually draining the reservoirs. In fact, because of evaporation, less water can be withdrawn than enters a reservoir system. Dams do not add water; they cause a net *loss* of water.

When interviewed in 1997, the elderly but energetic and lucid Floyd Dominy recalled why the bureau built Glen Canyon Dam: to protect the upper basin. Without Lake Powell, under the law of prior appropriation California would have been able to keep all the Colorado River water it used. In dry years, in order to meet its delivery obligations, without Lake Powell the upper basin might have to cut its own use and deplete its own reservoirs. In the CRSP hearings, Utah senator Arthur Watkins implied strongly that even without prior appropriation, Congress, heavily weighted toward the thirsty California bloc, might simply decline to authorize any reclamation projects in the upper basin, shutting them down as effectively as would new legislation.[29] According to Dominy, the fear of that possibility would have prevented the upper basin states from starting their own water projects and would have limited their progress. "They say that Glen Canyon isn't needed, you don't divert any water out of Lake Powell," Dominy explained. "Well . . . when you divert water through the mountains in Utah and Colorado . . . it's really out of Lake Powell because its the big storage."[30] As he also knew, without a big reservoir somewhere in the upper basin—and Glen Canyon would store far more than any other feasible site—its representatives might have voted against the CRSP.

The twenty-first-century drought provides the best test to date of whether Dominy and the supporters of Glen Canyon Dam were right—whether Lake Powell is truly needed to protect the upper basin against drought. During one of the worst droughts in five hundred years, would the upper basin have been able to meet its downstream obligation without Lake Powell? Remember that the Law of the River requires the four upper basin states "not to deplete" the flow of the Colorado below a ten-year rolling average of 7.5 MAF. Using a rolling average instead of an annual obligation buffers the extreme years substantially. We know that during the first eight years of the twenty-first-century drought, the upper basin has *not* curtailed its water use nor has the lower basin suffered a shortfall. Lake Powell released the water it was supposed to, even though it came at the expense of a full reservoir. Many have claimed that the ability of both basins to continue business as usual during the drought proves the wisdom of having built Lake Powell. But the real question is whether, without the reservoir, enough water would have flowed past Lee's Ferry to meet the Law of the River. The answer so far is unequivocally yes. As shown in the chart on the annual flow of the Colorado (p. 164), even now, the ten-year average inflow to Lake Powell has never dropped even to 10 MAF. Without the evaporation from Lake Powell, each year several hundred thousand acre-feet more would have flowed down to the lower basin. As I will explore in chapters 13 and 14, if flows remain below average, eventually the 7.5 MAF requirement is bound to be breached, but by then Glen Canyon Dam and Lake Powell will have even more serious problems.

Hydroelectric power revenues were to be the major benefit of Glen Canyon Dam. Indeed, as Paul Douglas showed, Glen Canyon was the only dam in the CRSP expected to make a profit. Power generation fell dramatically in the 1990s, when to protect the ecology of the Grand Canyon, dam operators at Glen Canyon were required to dampen the extreme swings in releases timed to generate and sell power at peak rates. The twenty-first-century drought has dropped the lake level and lowered the hydraulic head on the generators, reducing power production by one-third. Today, Glen Canyon generates about 8 percent of the power produced by bureau dams and 1 percent of the power on the western grid.[31] Hoover Dam can produce half again as much power as Glen Canyon; the Navajo Generating Station at Black Mesa and the Four Corners power plant both generate several times as much. Coal-fired plants in Utah and Wyoming also produce significant amounts of power. Thus events have overtaken Glen Canyon Dam and diminished power generation as a justification for it. The shortfall has caused the Western Area Power Administration to have to buy power on the spot market and pass the increased costs on to consumers.

At the hearings for Colorado River Storage Project, recreation received relatively little attention. The Southwest already had Lake Mead; it lay closer to the population centers of Southern California and as close to Phoenix as did Glen Canyon. What congressman would endorse spending hundreds of millions of dollars to build a second giant water playground at inaccessible Glen Canyon, especially since there was no easy way to calculate a benefit-cost ratio for recreation and demonstrate fiscal prudence? Only one person predicted what a mecca for vacationers Lake Powell would turn out to be. In 1959, Floyd Dominy forecast that over three million people a year would visit, at the time a seemingly absurd estimate. Yet visitation began to climb sharply in the early 1980s and by 1988 it had broken the three million mark, just as the old fox had predicted. But by 2006, visitation at the Glen Canyon National Recreation Area had dropped to 1.9 million and the little town of Page, Arizona, was beginning to look frayed.

———

In 1974, author Bruce Berger volunteered to drive Stewart Udall, who was in Aspen, Colorado, on a fund-raising trip, back to the airport. Berger took the occasion to pepper his captive audience of one with questions about the decision to dam Glen Canyon. "Couldn't they just measure the amount of water that passed Lee's Ferry and give the Upper Basin *credit* for their half of the water?" the author asked the former secretary. "Sure," Udall replied, "but Aspinall wanted dams. He said not to have dams would interfere with their bookkeeping system."[32] In other words, we have Glen Canyon Dam and Lake Powell because Aspinall and the Concrete Pyramid wanted them. The Bureau of Reclamation needed new dams to burnish its reputation and justify its funding and staff levels. Mega-construction companies, spawned at Hoover Dam, needed a continuous supply of new projects. Upper basin irrigators had for years felt shortchanged by the largesse the bureau had directed at their downstream competitors, especially the despised Californians. It was time the upper basin got its share, no matter that much of the land was too high and dry to earn a profit. Upper basin politicians had seen reclamation project after project benefit the lower basin. They wanted an insurance policy that would guarantee that when the upper basin needed Colorado River water, it would be able to take its full allotment. Utah's Senator Watkins explained:

> Inasmuch as water runs downhill, failure to authorize the upper
> Colorado River storage project under the same ground rules as have
> been afforded other reclamation projects will be tantamount to giving
> away the water and power resources of the upper basin States to the
> States of the lower Colorado River Basin and Mexico.[33]

Through the end of the 1960s, when their reign ended, the long tenure of western congressmen and senators, especially Aspinall and Hayden, had given them the power to do as they pleased in the West. Aspinall's successful insistence on his five western Colorado projects shows the extent of that power. Before power corrupts, it breeds arrogance. It leads to reclamation projects that make little or no economic sense and are defensible mainly as political acts. The Concrete Pyramid needed Glen Canyon and, while the conservationists turned their heads, seized it.

As Lake Powell continues to fall, the day may come when people view it as residents of Ventura County, California, view Matilija Dam and reservoir. Once seen as an essential solution to the water needs of California's Central Coast, sediment-clogged Matilija and its decrepit dam have become an expensive, dangerous, and insufferable problem. But at least Matilija is small enough so that its silt can be removed and the dam dismantled. Glen Canyon Dam is so vastly larger that to remove it would take decades, destroy the local environment, and bust any budget. Dams the size of Glen Canyon and Hoover are not going to fall at human volition. Instead they will stand until time and the river flowing dismantle them—as has happened to every obstacle in the path of the Colorado River. It may take centuries; it may take millennia—but assuredly it will happen.

As reservoirs gradually fill with silt, their capacity for holding water will decease and floods will begin to overtop the dams. Water cresting a dam will fall to a plunge pool at the base and erode the downstream toe of the dam, overbalancing it. The pressure of the wet sediment—twice as dense as water—or an earthquake, or a prolonged flood, will one day topple the off-balance dam. The freed sediment will quickly wash downstream, clogging the river channel and destroying any ecosystem that has survived. But soon the river will remove the extra sediment and return to an equilibrium balance between sediment supply and demand. Flowing water will gradually dissect the jumbled mass of broken concrete and wash it away. One day every trace of the dams and their reservoirs will be gone, a few exotic grains of concrete the only evidence of their one-time existence.

PART FOUR

River of Limits

THIRTEEN

Time Machines

WITH THE HISTORY OF THE Colorado River system of dams and reservoirs behind us, we can address two vital questions. First, what is the balance between supply and demand on the Colorado River today? Second, how is that balance likely to shift over the rest of the twenty-first century?

The chart on the following page shows the annual flow of the river at Lee's Ferry since measurements began in 1896, together with a ten-year running mean. The wet early years of the twentieth century, the ones that misled the Colorado River commissioners, are easy to spot. The chart also shows how the flow varies unpredictably from one year to the next and how, from 1930 to 1977, flows were lower than before or after that period. In fact, during that forty-eight-year stretch, the flows averaged only 13.6 MAF. Could that figure be closer to the river's long-term flow? How typical of the flow over the longer term were the twentieth-century discharges? To answer these kinds of questions, we would need to know the flow of the Colorado River year by year for the last several centuries. But surely that is impossible, for it would require a time machine.

At the end of the nineteenth century, a young astronomer named Andrew E. Douglass worked at the Lowell Observatory in Flagstaff,

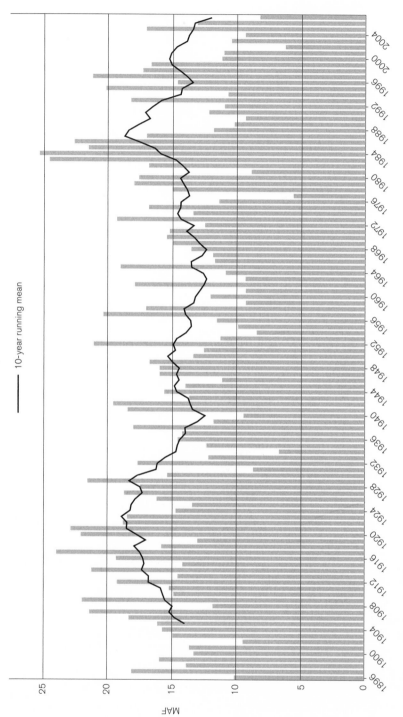

Figure 20. Annual flow of Colorado River at Lee's Ferry, with ten-year running mean (Adapted from Bureau of Reclamation)

Arizona.[1] The observatory took its name from Percival Lowell of Boston, who had sent Douglass west to find the ideal site for watching the heavens. Douglass focused his observations on Mars and, in the drawings and descriptions he sent back to Lowell, the Bostonian thought he saw linear features that resembled canals. That led Lowell to deduce that intelligent canal-builders live on Mars. Douglass disagreed and managed to distance himself from Lowell and one of the more embarrassing episodes in modern science. Their disagreement got Douglass fired, but he went on to invent a new branch of science that, unlike Lowell's loony notions, has stood the test of time; indeed, Douglass's time machine allows us to measure the flows of the Colorado River over many centuries.

As Douglass hiked the beautiful pine forests around Flagstaff, like countless individuals before him, he noticed the annular rings on tree stumps and broken limbs, some of the rings wide and some narrow. He speculated that sunspots might affect the earth's weather; if so, tree rings might record the eleven-year sunspot cycle. But as often happens, serendipity intervened and the scientist discovered something that he had not been looking for, something that had been there all along but that no one before him had discerned. Douglass saw that different tree stumps had the same pattern of wide and narrow rings—like a pair of matching bar codes, we would say today. He discovered that he could measure the width of the rings in a living tree by boring a small, pencil-sized hole and extracting a core. Douglass learned to match the pattern of bands from a living tree with those from a tree that had died but remained standing. The pattern in the standing, dead tree overlapped those of a dead tree lying on the ground, those in turn overlapped the rings in a timber used in an ancient dwelling, and so on until Douglass had identified a unique sequence of overlapping time markers stretching back hundreds of years. By counting the rings inward, he could determine the year in which each formed.

Trees grow by laying down radial bands of cells on the outside of those laid down earlier. Where temperature and moisture vary with the seasons, as in the Southwest, trees produce larger cells in the spring and smaller ones later in the year. At the junction of the two regions of different cell density, a distinct, darker ring often forms, as we see on the polished tree trunks in science museums. The height of a tree turns out to depend primarily on climate and secondarily on the quality of the soil, the slope, competition from other trees, and so forth. But the width of the rings depends largely on annual weather, with wider rings forming in wetter years and conversely. Thus the rings provide not only a chronometer, but an ancient weather gauge. Douglass had invented dendrochronology.

One of his major achievements was to use tree rings to date the famous Anasazi site at Pueblo Bonita in Chaco Canyon, New Mexico. Applying his methods to the wood that framed the ancient rooms, Douglass dated the ruins to the sixth century A.D. Clusters of narrow bands told him that drought had forced the Anasazi to migrate.

At the Laboratory of Tree-Ring Research at the University of Arizona, which he founded, Douglass and his colleagues soon discovered that trees sometimes generate false rings; at other times, a ring that ought to be present is missing. Thus simply counting rings from a single tree and matching them up with rings from other trees is not enough. Modern tree-ring scientists draw cores from several trees at many sites, gathering so much data as to require sophisticated computer analysis and complex statistical techniques.

Though the work requires painstaking care, it is easy to see in principal how scientists use overlapping tree-ring patterns to work backward from living trees to date ancient artifacts and how the width of the rings allows them to detect past wet and dry periods. But they also use tree rings to gauge the prehistoric flow of the Colorado River—how are they able to do that?

By the 1970s, the tree-ring specialists had two sets of data. They had the recorded flow of the Colorado River since 1896, which must correlate with annual precipitation in the river basin. For the same period, they had tree-ring data from living trees at many different sites on the Colorado Plateau. Douglass and others had shown that the width of tree rings also correlates with annual precipitation. In effect, the dendrochronologists removed the common factor, annual precipitation, and correlated river flow directly with the tree-ring patterns. They applied their method to the oldest trees on the Colorado Plateau, some of which lived more than one thousand years ago.

In 1976, Charles Stockton of the Laboratory of Tree-Ring Research and Gordon Jacoby of UCLA extended the tree-ring record in the Colorado River basin back nearly 450 years.[2] The data from the two reconstructions in which they had the most confidence gave a mean flow for the Colorado River of 13.5 MAF. Stockton and Jacoby found that the periods from 1564 through 1600 and from 1868 through 1892 were drier than any in the twentieth-century record. Conversely, the period from 1906 to 1930, during which the Colorado River commissioners met, "was the greatest and longest high-flow period within the last 450 years." Ever since, the Stockton-Jacoby report has hung like a sword of Damocles over western water planning.

Figure 21. Tree-ring scientist Connie Woodhouse (Photo by Kurt Chowanski)

In 2000, a group of scientists reanalyzed Stockton and Jacoby's data. They confirmed the timing of the droughts, but found that the dry spells were even more severe than Stockton and Jacoby had thought. The modern reanalysis of the older data gave a mean flow of only 13.2 MAF.[3] Could the Colorado River really carry that little water?

In 2006, University of Arizona scientists Connie Woodhouse, Stephen Gray, and David Meko reported their tree-ring study of the basin.[4] Compared with Stockton and Jacoby, these scientists sampled different trees, took more samples from each, had thirty more years of gauged flows for their calibration, and used more sophisticated methods. Four different ways of reconstructing the tree-ring data gave a mean flow of 14.6 MAF. Why the earlier studies gave lower means is still unresolved, suggesting we may have not heard the last word. But for the time being, the average of these most recent results, 14.6 MAF, can stand as the best estimate of the long-term natural flow of the Colorado River before modern global warming depletes it. To use a higher number—say, the average for the atypically wet twentieth century— as the basis for long-term planning would be unscientific and irresponsible.

The 2006 paper by Woodhouse and colleagues fully corroborates the earlier findings that dry periods in the Southwest can appear suddenly and last for years, even for a decade or two. The new study detected several droughts that were as bad as any in the twentieth and twenty-first centuries and one, between 1844 and 1848, that was worse. During that period, the flow of the Colorado River averaged only 9.6 MAF, two-thirds of its centuries-long average. A subsequent paper by Meko, Woodhouse, and colleagues extended the tree-ring record in the Colorado Plateau back to A.D. 762.[5] Its most startling finding was that during A.D. 1130–1154, the mean flow was less than 85 percent of the twentieth-century mean and that for a longer period of six decades, high annual flows were absent.

In a river as erratic as the Colorado, the deep, long-lasting droughts that the tree-ring studies reveal may matter as much as the long-term flow of the river. Even in a game with 50–50 odds, a gambler can have an unlucky streak, go broke, and have to quit the game before the eventual long-term odds catch up. In the same way, a long drought can drop the flow of a river so low that one or more of its reservoirs falls to dead pool. Wet years may one day return and restore the long-term average flow, but just as the gambler who runs out of money has to leave the game, after a reservoir reaches dead pool, things are different. To try to meet its downstream obligations, at dead pool Glen Canyon Dam would have to release through its outlet works every drop of water that entered Lake Powell, returning control of the river to Nature. Power generation would have ended. In the case of Lake Powell, only two boat ramps would reach the water and a bathtub ring of skyscraper height would band the lake. The Law of the River would break down and a legal civil war would ensue.

Using the figure of 14.6 MAF from the most recent tree-ring study, we can calculate the balance between supply and demand on the Colorado River today. By adding the inflow from the tributaries between Glen Canyon and the Mexican border, about 1.2 MAF, we obtain a total supply of 15.8 MAF. Consumption and reservoir evaporation make up the demand side. The Bureau of Reclamation projects current upper basin consumption at

4.5 MAF; Hoover Dam releases 7.5 MAF to downstream users and sends 1.5 MAF on to Mexico.[6] The amount of water evaporated from the two large reservoirs depends on how full each is; let us use the historic average for Lake Powell of 0.56 MAF and for warmer Lake Mead of 0.80 MAF. Evaporation below Hoover Dam and other losses cost the system about 1 MAF, bringing total demand to 15.9 MAF. One can adjust these numbers up or down a bit, but not enough to change the conclusion that *today*—not sometime in the future after global warming has reduced supply—the surpluses on which the basin states have long counted are gone. As Mister Micawber described the family balance sheet, "Annual income twenty pounds, annual expenditure nineteen, nineteen six, result happiness. Annual income twenty pounds, annual expenditure twenty pounds ought and six, result misery."[7]

Regardless of the state of the Colorado River today, every city, county, and state in the West plans to consume more water. In defiance of logic and limits, the driest states have become the fastest growing. According to the United States Census Bureau, Utah, the most arid state, is the fifth fastest-growing. Colorado is fourteenth and New Mexico twenty-sixth. In the lower basin, Nevada, Arizona, and California are first, second, and thirteenth, respectively. Clark County, Nevada, home to Las Vegas, is the fastest-growing county in the nation. Every state and major city in the West plans to add more people and to provide them with housing, electricity, and water. But if there is no surplus in the Colorado River system now, from where will the water come to meet the increased demand?

Ominously, during the first eight years of this new century, including a wet 2005, the Colorado River has run at only 60 percent of its long-term flow. One reason for the severity of the current drought, as revealed in the chart on the next page on annual temperatures in the West over the past century, is that it is occurring under higher temperatures than during any of the previous twentieth-century droughts. The chart shows that the droughts of the 1950s and 1970s took place at temperatures at least two degrees lower than today. Higher temperatures increase evaporation from snow and thus reduce snowmelt. They also increase absorption of rain and snowmelt by dry soils. Both effects reduce runoff and the amount of water in streams and reservoirs. Are the temperature increases and other changes observed in the West mainly natural, or mainly caused by humans? A group of scientists led by Tim Barnett of the Scripps

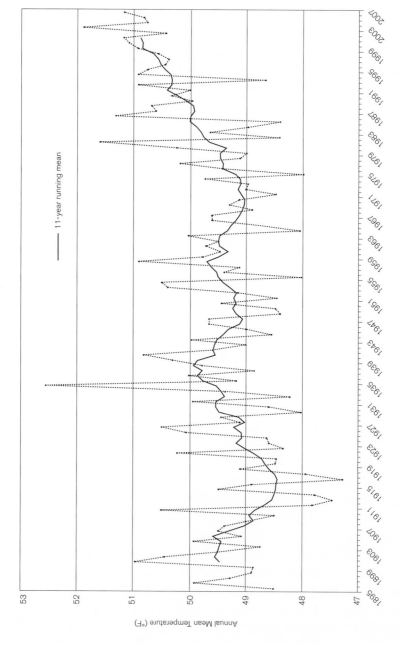

Figure 22. Annual western U.S. temperatures, with eleven-year running mean (Adapted from Dr. Kelly Redmond and the Western Regional Climate Center)

Institute of Oceanography answered that question in a paper in *Science* magazine in early 2008: "Up to 60 percent of the climate-related trends of river flow, winter air temperature and snow pack between 1950–1999 are . . . due to human-caused climate changes from greenhouse gases and aerosols." They forecast "a coming crisis in water supply for the western United States."[8]

The West has been growing warmer—but so has the entire planet. During the twentieth century, mean earth temperature rose by $1.1 \pm 0.4°$F. But three-fourths of the increase occurred in the last three decades of the century.[9] According to the latest report from the Intergovernmental Panel on Climate Change (IPCC), temperatures in the Northern Hemisphere during the second half of the twentieth century were higher than "during any other 50-year period in the last 500 years . . . and [were] likely the highest in at least the past 1,300 years."[10]

The statistics pile atop each other: the 1990s were the warmest decade on record; 1998 was the warmest year ever—until 2005. From January through June of 2006, the average temperature for the lower forty-eight states was 3.4°F higher than the twentieth-century average. July 2006 broke 2,300 daily United States temperature records.[11] Each of the warmest twelve months on record in the United States has occurred since 1997—and January 2007 broke the monthly record by a wide margin. In the contiguous United States, 2006 was the second hottest year on record, missing the record set in 1998 by only 0.1°F.[12]

When global temperatures over the last millennium are plotted, they resemble a hockey stick resting on its side, with the blade corresponding to the sharp rise during the twentieth century. Some have charged that the hockey stick is an artifact of the way the data were collected and analyzed. To assess that claim, the National Academy of Sciences convened a panel of distinguished scholars. After reviewing the evidence, the panel had a "high level of confidence that global mean surface temperature was higher during the last few decades of the 20th century than during any comparable period during the preceding four centuries."[13] The hockey stick is real.

The question is not whether the earth has warmed, but why? The scientific consensus is that the cause is the increase in atmospheric carbon dioxide and other greenhouse gases, which absorb heat and trap it near the earth. In one of the most prescient predictions in science, in 1896, long before temperatures had started to rise and long before the terms "greenhouse gas" and "global warming" had been coined, Swedish chemist

Svante Arrhenius predicted the very rise that we now observe. Based on the knowledge that carbon dioxide molecules trap heat, Arrhenius calculated that if atmospheric carbon dioxide levels were to double, global temperatures would rise between 7 and 11°F. More than a century later, with vastly more information, the IPCC forecasts that by 2100, temperatures will rise between 2.5 and 10.5°F, overlapping the range the Swedish chemist forecast long ago. Arrhenius thought it might take three thousand years for carbon dioxide levels to double, but sadly that is one forecast that he got wrong.

The greenhouse effect arises because the sun's rays, which are of short wavelength, pass through the atmosphere nearly unimpeded and are absorbed at the earth's surface. But the radiation that our planet re-releases lies in the longer wavelength infrared band, which greenhouse gases in the atmosphere strongly absorb. The effect traps heat and makes the earth about fifty degrees warmer than it would otherwise be. Without greenhouse warming, our planet would be uninhabitable. These are incontrovertible scientific facts.

The principal greenhouse gases are water itself, carbon dioxide, methane, and certain other oxides and fluorocarbons. Since these gases have kept the earth warm enough for us to inhabit, logic dictates that as their concentrations increase, the earth must grow warmer. And concentrations of greenhouse gases have increased. In the late 1950s, scientist Charles Keeling began to measure carbon dioxide concentrations at the scientific station atop two-mile-high Mauna Loa in Hawaii. Since Keeling began his measurements in 1958, carbon dioxide levels have climbed steadily from 315.7 parts per million to 386.5 ppm in 2007. To estimate carbon dioxide levels before Keeling began his precise measurements, scientists use cores extracted from polar ice sheets. They date the annual levels in the ice and extract gases from the bubbles. Measurements from Antarctic ice cores show that prior to the industrial revolution, earth's atmosphere contained about 280 ppm of carbon dioxide. Since 1880, when temperature records begin, though the correlation is not perfect, carbon dioxide levels and temperature have risen together. One reason for the lack of exact correlation is that temperature also rises and falls due to natural causes. More heat may escape the earth's core or the intensity of the sun's rays may increase—both make the earth warmer. Volcanic eruptions like that of Mount Pinatubo in the Philippines in 1991 insert aerosols into the atmosphere that absorb radiation and lower global temperatures. We can be sure that one warm spell had a natural cause: between about

A.D. 900 and 1300, and thus before greenhouse gas levels had increased, the earth underwent the Medieval Warm Period. Scientists are not sure why, but they do know that the warming was much less than the planet is currently experiencing. When climate scientists use both natural and man-made factors, their computer models reproduce observed temperatures almost exactly. There are two possible explanations: cause and effect, or coincidence.

———

Many kinds of evidence convince scientists that the correlation between rising greenhouse gas levels and rising temperature is not coincidence. Here are five examples of the evidence. First, in natural cycles of heating and cooling, such as the El Niño–La Niña pairing, some regions of the earth warm while others cool. But according to the IPCC, twentieth-century temperatures have risen on every continent except possibly on Antarctica, where we do not have enough data to know. Second, the troposphere, the atmospheric layer nearest the earth's surface, has warmed while the layer above it, the stratosphere, has cooled. If the source of global warming were the sun rather than the earth, the stratosphere should have heated at least as much as the troposphere. Third, not only have daytime temperatures risen, so have minimum nighttime temperatures. Even after the sun sets, something holds heat close to the earth. Could that something be the concrete and asphalt in urban heat islands? No—the oceans have heated in parallel with the land and therefore the warming is global, not continental.[14] Fourth, the solar sunspot cycle can cause temperatures at the surface of the earth to increase, but the maximum that has been measured since the late 1800s is only 0.2°F, a fraction of the observed rise. Fifth, in the second half of the twentieth century, the world's oceans heated by 0.7°F. As *Newsweek* points out in an excellent article, that does not seem like much, but multiply it by the vast volume of the earth's oceans. During the same period, the sun's energy output rose by less than 0.1 percent.[15]

This kind of evidence has convinced the vast majority of scientists that greenhouse gases are the principal cause of global warming. Contrary to the impression given by some pundits and politicians, among scientists there is next to no controversy. Author and historian of science Naomi Oreskes reviewed the abstracts of 928 scientific papers that discussed "global climate change" between 1993 and 2005.[16] How many tried to refute the theory that man-made emissions of greenhouse gases are a major cause of global warming? Not one. Some scientists have expressed skepticism before congressional committees, in the press, and on talk shows, but evidently over the twelve-year period that Oreskes reviewed, none published their doubts in peer-reviewed professional journals.

In February 2007, the Intergovernmental Panel on Climate Change issued the first of four reports in its latest assessment of climate change.[17] Over 2,500 scientists from more than 130 countries participated. The report, edited with the "assistance" of government officials from around the world, concluded that there is a very high confidence—at least 90 percent—that most of the observed increase in global temperatures is due to man-made greenhouse gases. Comparison of the scientists' final draft and the edited IPCC report shows dozens of instances in which the government officials watered down the scientists' conclusions—and not a single instance where the bureaucrats strengthened the wording. If out of this process, a conclusion can be stated with "very high confidence," only conspiracy theorists could ignore it.

Some scientists have recorded their concern that the latest IPCC report understates the danger from climate change.[18] One reason that reality may be worse than the forecasts is that global warming has insidious feedbacks that are difficult to model. For example, warmer temperatures increase evaporation; warmer air can hold more water. But water vapor is itself a heat-absorbing greenhouse gas. The more water vapor in the air, the more heat it traps, the more the earth warms, the more water evaporates, the more water vapor in the air, and so on. The IPCC report says that "water vapor changes represent the largest feedback affecting climate sensitivity."[19]

Ice reflects about 80 percent of solar radiation; ocean water absorbs about 90 percent. As the earth warms and ice melts, the water absorbs much more heat than did the ice. The effect causes more ice to melt, the resulting water then absorbs still more heat, and so on. On land, shrinking ice and snowpack expose soil, which absorbs more heat, causing more warming and more shrinkage.

Permafrost, or frozen ground, covers about 20 percent of the earth's land mass, mainly in the Arctic. Methane does not last as long in the atmosphere as carbon dioxide, but while there it traps twenty times the heat. Methane levels have also climbed. As permafrost warms and the thawing vegetation releases methane, the earth warms more, causing more permafrost to thaw, releasing more methane, and so on. A team of American and Russian scientists found that Siberian lakes are emitting methane five times faster than previously estimated.[20]

Since the 1980s, the number of western forest fires has increased fourfold. Fires destroy trees that otherwise would remove carbon dioxide from the atmosphere. Their absence causes more warming, which leads to more fires, and so on. Warmer temperatures stimulate bark beetles, which kill more trees, producing another feedback. In some Canadian forests, bark beetle populations have exploded, killing more trees than logging or wildfires.[21]

Climate change produces other positive feedbacks—those that make things worse—as well as a few negative ones—those that make things better. How the feedbacks will interact and reinforce each other, or not, no one really knows. Perhaps the feedbacks are the reason that the reality of climate change is turning out to be worse than the IPCC projected when it made its first forecasts in 1990.[22] The carbon dioxide levels forecast then have turned out to be accurate, but temperature and sea level have risen faster than the projections. For example, sea level has risen faster in the last twenty years than in any similar period in the preceding 115 years.

Financial and insurance companies, who will be held to account by shareholders and go broke if they get it wrong, are not to be found among those denying global warming. To decide how seriously to take global warming, read their hard-nosed reports. A summary by the United Nations Environmental Finance Initiative says that "climate change is now certain" and projects that by 2040, disaster losses due to global warming could exceed $1 trillion each year.[23] The British insurance industry projects that by 2080, climate change will increase the cost of hurricane damage in the United States—not including damage from wind and other storms—by three-quarters, to a total of $100 to $150 billion dollars annually. That is the equivalent of a Hurricane Katrina—each year.[24]

It is already too late to prevent the earth from warming significantly in the twenty-first century. Suppose that the main carbon dioxide emitters—the United States, China, and India—were to stabilize emissions at year 2000 levels. Because the world's oceans absorb and emit carbon dioxide on a time scale of hundreds of years, global temperatures would still rise by at least another degree. The IPCC report gives fair warning: "Anthropogenic warming and sea level rise would continue for centuries due to the timescales associated with climate processes and feedbacks, even if greenhouse gas concentrations were to be stabilized."[25] Global temperatures are rising, will go on rising, and will have their worst effects in the driest regions.

—

Groups of experts, with their specialized training and terminology, may have trouble explaining their findings in a way the public can easily comprehend. Climate change scientists are no exception. They apply a dozen or more supercomputer models to the several IPCC emission scenarios, each assuming a different degree of increase in greenhouse gas levels. The result is often a range of projected effects, tempting laymen to focus on the low end, the one that will require us to do the least.

To more dramatically illustrate the possible danger, one group of scientists has modeled the world in 2100.[26] They used three climate-change models, adjusting them until they replicated the known temperatures between 1960 and 1990. Then they applied the models to two of the IPCC emission scenarios (labeled A and B). Scenario B2 projects a rise in atmospheric carbon dioxide to 620 ppm by 2100; A2 projects a rise to 860 ppm. The scientists ran the three models using both scenarios and computed overall averages. To present the results they ask us to recall the hottest year we have known in the last twenty years. For most of us, that is likely to have been 2006. Next they ask, by 2100, how often will a year that hot occur? If the answer were "one year out of twenty," the scientists would foresee a world no hotter than today. "Ten years out of twenty" would indicate that by 2100, one out of every two years will be as hot as the hottest year of the last twenty. But the answer from the scientists is "twenty years out of twenty." *By 2100, the hottest year in twenty today will be the norm.* The scientists illustrated their results using a colored map of the world. The greater the number of hot years in twenty, the redder the scale. In 2100, their map shows a crimson earth.

Next the scientists turned to precipitation. By 2100, how often will today's wettest year in twenty occur? We might hope that precipitation will increase enough to mitigate the effects of higher temperatures. Instead, by 2100, in most regions of the earth the wettest year in twenty occurs with about the same frequency as today. With a few exceptions, precipitation increases most at the highest latitudes, where it will do the least good. Some regions, like Brazil and southwestern Africa, will not only grow much hotter, they will become much drier. In those areas, the maps spell catastrophe.

For the contiguous United States, the two maps show that by 2100 the number of hot years will increase by several times the number of wet years, a recipe for depleted rivers and reservoirs. Nowhere in the country will this be more true than in the Colorado River basin and the Southwest. Indeed, changes there are already well under way. As shown in the preceding chart, the average temperature of the western United States has climbed steadily, most noticeably since 1950. In each of the four major river basins in the West—the Colorado, Columbia, Missouri, and Rio Grande—the years between 2000 and 2004 were hotter than the historical average by at least 1.5°F. The Colorado River basin was 2.1°F hotter.[27] Since 1986, longer and hotter summers have led to six times as many acres burned in wildfires. The length of the wildfire season increased by more than two months and the average duration of large fires rose from 7.5 days to 37 days.[28] Scientists attribute these increases to rising summer temperatures and mountain snowpacks that melted earlier. The sooner snowpacks melt, the sooner humidity falls, the sooner soils dry out, and the sooner the fire season starts and the longer it lasts.

Even a slight reduction in snowpack spells trouble. So much of the water in the Colorado River basin evaporates that nearly 90 percent of the water in streams must come from a virtual reservoir: the Rocky Mountain snowfields.[29] Rising temperatures cause more winter precipitation to fall as rain, reducing snowpack levels at the outset. These reductions have already begun. Between 1950 and the 1990s, even though overall precipitation rose slightly, the volume of Rocky Mountain snowpack declined by 16 percent. In each of the four western river basins cited above, most years since 1990 have had less snowpack than the historical average.

The higher the temperature, the earlier snowpacks start to melt. One study found that for 279 streamflow gauges, both snowmelt and peak streamflow across much of western North America began ten to thirty days earlier than previously.[30] Earlier melting sends water downstream in the spring, before cities and fields can use it. If reservoirs have room, they can store the earlier-arriving water; if not, they will overflow. A crucial question is whether global warming will deplete western reservoirs enough so that they will be able to contain the earlier-arriving and larger spring snowmelt. Lakes Powell and Mead will have room, but smaller reservoirs in other states may not. California's governor has proposed additional reservoirs to hold the earlier snowmelt rush. Should the state spend billions building new reservoirs, when existing ones are partially empty and when global warming is going to deplete them further? The question is hard to answer, but money spent on conservation measures instead of dams would be sure to work and provide lasting benefits.

One of the IPCC emission scenarios projects twenty-first-century global mean temperature increases ranging between 3.6 and 9.7°F.[31] What might such increases mean for the people of the West? If the actual rise is toward the high end of that range, Seattle will become as warm as Sacramento today, Portland as warm as Los Angeles, Missoula as warm as Denver, and Phoenix as warm as Death Valley.[32] Like Major Powell and his crew as they launched into the unexplored canyons of the Colorado, the West has already entered new territory, with no maps and little knowledge of what lies ahead.

FOURTEEN

A New Climatology

WE KNOW THAT TEMPERATURES in the West, as in the rest of the globe, have been rising. How might they go on rising and what might be the effect on the Colorado River system of dams and reservoirs? To answer these questions and forecast temperature, precipitation, and runoff, climate scientists use complex supercomputer models. Like the tree-ring specialists, the modelers first adjust their programs until they replicate the measured temperatures during the twentieth century. One modeling exercise, published in 2004, was able to match the historic flow at Imperial Dam, near the Mexican border, with an accuracy of 99 percent.[1]

The climate models collectively portray a twenty-first-century Colorado River basin that is hotter but no wetter. Warmer temperatures not only reduce the size of the snowpack and cause it to melt sooner, they increase evaporation from snow and soils, as well as transpiration from plant leaves (together called evapotranspiration). Setting aside the relatively small percentage of rainfall that sinks into the soil, runoff is essentially the difference between precipitation and evapotranspiration. The following table shows that between 1950 and 2000, an average of 354 millimeters of precipitation fell on the Colorado River basin. But 309 mm of the total, or 87 percent, evaporated or transpired. Because runoff in the West is the difference between two large and similar numbers, a small change in either precipitation or evapotranspiration causes a multiplied

TABLE I
Increased Evaporation in Colorado River Basin Causes Disproportionate
Decrease in Runoff

	Precipitation (mm/yr)	Evapotranspiration (mm/yr)	Runoff (=difference)
1950–2000 observed	354	309	45
Change one factor	354 (0%)	315.2 (+2%)	38.8 (−13.7%)
Change both factors	350.5 (−1%)	315.2 (+2%)	35.3 (−22%)

Niklas S. Christensen and Dennis P. Lettenmaier, "A Multimodel Ensemble Approach to Assessment of Climate Change Impacts on the Hydrology and Water Resources of the Colorado River Basin," *Hydrology and Earth System Sciences Discussions* 3 (2006): 3727–3770.

percentage change in runoff. As shown in the table, if precipitation remains constant while evapotranspiration increases by just 2 percent, runoff declines by nearly 14 percent. If precipitation falls by 1 percent and evapotranspiration increases by the same 2 percent, runoff drops by 22 percent. Thus not only is the water balance in the Colorado River basin poised on a razor's edge between supply and demand, even a tiny increase in evapotranspiration causes a multiplied and dangerous decrease in runoff.

In the last few years, several teams of scientists have published their runoff forecasts for the Colorado River basin. Each projects less water in the Colorado River, but they differ on how much the reduction will be. One study used a single climate-change model to forecast that by 2050, runoff in the basin will decline by 18 percent.[2] In a later paper, a subset of the same authors used eleven different climate-change models to project a decline of 6 percent by mid-century.[3] Another study used an ensemble of twelve models to forecast a decline between 10 and 30 percent.[4] Still another found that western droughts may last an average of twelve years, at least twice the duration of twentieth-century droughts. According to this study, climate change could cut the flow of the Colorado River at Lee's Ferry by a crippling 60 percent.[5] Just before Easter 2007, a paper in *Science* magazine made front-page headlines in western newspapers.[6] The paper reported that "the levels of aridity of the recent multiyear drought, or the Dust Bowl and 1950s droughts, will, within the coming years to decades, become the new climatology of the American southwest." For comparison, in the early Dust Bowl years and the mid-1950s, flows averaged 10–11 MAF. So far in the twenty-first century through September 2007, they have averaged about 11 MAF. In the West, the new climatology may already have begun. Finally, consider the second of four reports issued by the IPCC in 2007.[7] It concludes that in some dry, mid-latitude regions, runoff will fall by 10 to 30 percent over the next fifty years. Assimilating these different studies, one would certainly be justified in assuming that global warming will reduce runoff in the Colorado River by 20 percent by mid-century. To assume no reduction would be imprudent or worse.

—

So far in its model forecasts for the Colorado River basin, the Bureau of Reclamation has chosen not use the scientific evidence from tree-ring studies and climate change. To forecast the future, the bureau uses the record of measured flows starting in 1906, even though tree-ring research shows the twentieth century was the wettest of the last five centuries. That is like assuming that your income for the rest of your life, including your retirement and twilight years, will equal your all-time high income. That kind of planning would lead you straight to the poor house.

How does the bureau justify ignoring tree rings and climate change? The agency answered in a 2004 report describing its Lower Colorado River Multi-Species Conservation Program and has since provided more information. The goal of the LCRMSCP is to allow the same or more water consumption from the lower Colorado River without further endangering the twenty-six protected species that live there. To meet these opposite objectives and to avoid having environmental concerns limit consumption, the bureau's forecasts need to show that the lower Colorado River will contain plenty of water for the foreseeable future.

We can use the level of Lake Mead as proxy for the amount of water in the lower Colorado. If the reservoir stands well above its power pool, Hoover Dam will be able to release enough water for people and species downstream. The bureau's model projects the possible future elevations of the reservoir, assigning each a probability of occurrence. The bureau's median outcome, and therefore its most likely, is that in 2050 Lake Mead will stand a comfortable 54 feet above its power pool.[8] On the river, business as usual can continue.

Some who attended the public hearings on the LCRMSCP urged the bureau to "predict the effect of climate change on actual flows in the Colorado River." In its published report, the agency demurred:

Reclamation believes that use of the actual data recorded over the past century provides the best basis for ongoing Colorado River management activities. . . . If Reclamation were to use a different modeling approach . . . it would conflict with all of the other Colorado River management actions and analyses that Reclamation has taken and is currently taking. Attempting to predict global changes in climate, shifts in demographic patterns, and other factors affecting Colorado River hydrology are far more speculative than Reclamation's reliance on actual annual hydrologic data.[9]

Three years later, well into a drought of historic proportions, the secretary of the Interior wisely asked the bureau to report on the likelihood that the basin states might be unable to meet their obligations to deliver Colorado River water. As in the LCRMSCP report, again the agency used the wet twentieth century as its baseline for future planning; again it ignored climate change. Its shortage analysis foresaw no shortages.

The bureau's draft report appeared within days of two others. One was the first IPCC report of 2007, in which the international climate panel not only concluded that temperatures are rising, but also—with 90 percent certainty—that humans are to blame. The other was a report from the National Academy of Sciences–National Research Council (NRC) entitled *Colorado River Basin Water Management: Evaluating and Adjusting to Hydroclimatic Variability.*[10] The NRC report summed up that "the combination of limited Colorado River water supplies, rapidly increasing populations and water demands, warmer regional temperatures, and the specter of recurrent drought point to a future in which the potential for conflict among existing and prospective new users will prove endemic."[11]

The bureau's final environmental impact statement for how to operate the Colorado River system in the face of prolonged drought appeared in November 2007. In contrast to previous bureau reports, this latest summary devotes several pages to climate change, asserting that it has been a subject "for discussion for many years." While recognizing that the IPCC and the climate scientists cited above project runoff reductions for the Colorado River basin, though expressed in different words, the bureau's position has not changed:

Additional research is both needed and warranted to quantify the uncertainty of these estimates in terms of the actual uncertainty in the climate response as well as the uncertainty due to differences in methodological approaches and model biases in order to better understand the risks of current and future water resource management decisions.

The experienced reader can easily translate. Sure enough, a few pages later: "Based on the current inability to precisely project future impacts of climate change . . . this final FEIS [Final Environmental Impact Statement] is based on the resampled historical record."[12] It is peculiar that the bureau would demand precision as a precondition for taking climate change into account, when its own model outcomes are presented not as precise river flows but as a wide range of probabilities. In any case, why could not the bureau perform, say, two sets of projections: one without a climate change factor and the other assuming some modest runoff reduction by 2050.

Using the bureau's preferred method of operating the Colorado River system, its modeling projects that Lake Powell will gain water between now and about 2030, after which the reservoir level will decline just slightly. The fiftieth percentile projection and therefore the most probable is that in 2050 Lake Powell will stand at about 3,660 feet, barely forty feet below full pool. The possibility of the reservoir reaching dead pool is so remote that the level does not appear on the bureau's charts. The bureau will recommend to the secretary of the Interior that these guidelines remain in effect until 2025.

All of this has a certain through-the-looking-glass quality. What could explain the bureau's reluctance to base its models on the best science? Perhaps only an insider can know, but using only David Brower's ninth-grade arithmetic, we can find a clue. In its modeling, the bureau uses the measured Colorado River flows from 1906 through 2005. The flows average 15.0 MAF—400,000 acre-feet more than the tree-ring average of 14.6 MAF. A second discrepancy arises when we take into account the runoff reduction that global warming could cause. The studies summarized above do not agree on how large that reduction might be, but they certainly justify the assumption that by 2050 global warming will reduce runoff in the Colorado River by 10 percent, or 1.45 MAF, annually. The sum of the two discrepancies thus comes to nearly 2 MAF a year. In thirteen years, that accumulates to 26 MAF, roughly the present volume of the two reservoirs above their dead pool levels. Were the bureau to include tree-ring and climate change science in its forecasts, it would have to project the reasonable possibility that within two or three decades, the Colorado River system of dams and reservoirs could fail.

In a paper titled "When Will Lake Mead Go Dry?" scientists Tim Barnett and David Pierce predict just that. Using their own method of analysis, they conclude that unless the bureau changes the way it operates the Colorado River system, there is a 50 percent chance that both

Lake Powell and Lake Mead will fall below their power pools by 2017 and that both will reach dead pool by 2021. A probability of 50 percent means that there is an equal chance the reservoirs could fall to dead pool later— *or sooner.* Perhaps the bureau's attitude reflects policy set at a higher level. Consider the reaction of the White House Council on Environmental Quality to the fourth and final IPCC report of 2007. By the time this synthesis document appeared in mid-November, the IPCC had shared the Nobel Peace Prize with former vice-president Al Gore. Throughout the year, evidence that humans had caused most climate change had grown stronger and a shift in public attitude appeared to be under way. Possibly for such reasons, the IPCC now had the courage of its convictions, expressing "very high confidence" that humans have caused global warming. Reductions in greenhouse gases need to begin now, the report urged, otherwise the world was in for a global climate disaster that could cause the extinction of 20 to 30 percent of world species. Melting of the Arctic ice sheets, not taken fully into account in the earlier reports, could cause sea level to rise much more rapidly than previously thought.

As before, delegates from 130 countries scoured the draft report. According to one observer, the United States delegation tried to delete the section titled "Reasons for Concern," which listed the possible negative effects of climate change. But this time the scientists made their language stick.[13] When asked at a news conference, "What is dangerous climate change from the perspective of the White House," the chair of the White House council responded, "We don't have a view on that."[14]

In this day of desktop computing it was inevitable that other organizations would build their own easy-to-use models of the Colorado River system and make them accessible over the Internet. The model from Arizona State University's Decision Center for a Desert City is for those who wish to forecast water supplies in central Arizona.[15] The Colorado River Open Source Simulator (CROSS) model, which climate scientist Niklas Christensen built for the environmental organization Living Rivers, is a simplified, spreadsheet-based version of the bureau's model.[16] Using the same assumptions as the bureau's model, CROSS gives virtually identical results. The reader can use either model and vary the input assumptions at will. CROSS gives more detailed results and is the model used in the rest of this chapter.

In running CROSS, I make the following assumptions:

· Lake Powell and Lake Mead start with the volume each had on 1 October 2007: 49 percent of capacity.

· The natural flow of the Colorado River, before global warming reduces it, is the five-hundred-year tree-ring average of 14.6 MAF.

· By 2050, global warming reduces runoff in the Colorado River basin by 10 percent.

· Upper basin consumption rises to 5.4 MAF by 2060, the same assumption the Bureau of Reclamation makes.

· The Law of the River requires Glen Canyon Dam to release 8.23 MAF each year. This obligation receives the highest priority in the model and must take place even if the upper basin has to deplete its own reservoirs and reduce its consumption.

· Lake Powell is allowed to fall to dead pool. But in order to simulate protecting Las Vegas's water supply intakes, when Lake Mead falls to 1,083 feet, the model reduces delivery obligations from Hoover Dam. The model does not permit Lake Mead to fall below 1,000 feet, even if that requires further reductions in downstream deliveries to United States and Mexican users. The effect of these rules is that, if necessary, Lake Powell will be sacrificed in favor of Lake Mead.

The user can choose any year from 1986 to 2006 as the start year; the model then inserts the flows from the historic record from that year forward and performs its calculations. Let us choose as the start year 1922, when the Colorado River commissioners signed the compact. The wet years were about to end then and the Dust Bowl drought was still several years off. The following chart shows the amount of live storage in Lakes Powell and Mead combined using these assumptions. Also shown is the combined live storage using the Bureau of Reclamation's assumptions. In a given year, the two sets of assumptions produce results that differ by 20 MAF or more. Assumptions based on science show that within three decades, the Colorado River system is likely to have enough water for only one reservoir; the bureau's assumptions show Lakes Powell and Mead holding plenty of water for the rest of the century. Not shown on the chart is the level of Lake Powell itself: using science-based assumptions and favoring Lake Mead, the model shows Lake Powell falling to dead pool in the early 2040s.

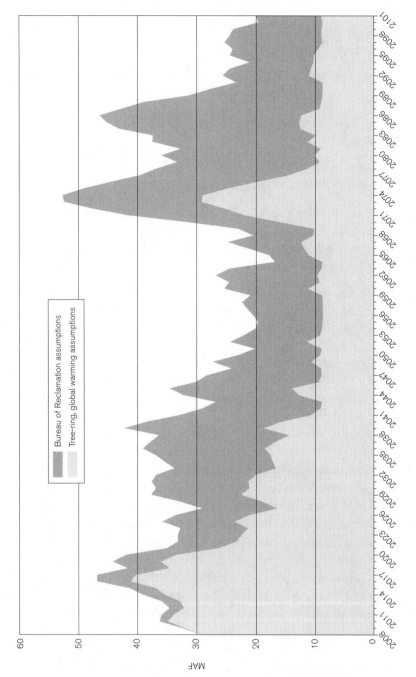

Figure 23. Divergent models of the future of the Colorado River system in the twenty-first century

Not shown on the chart but forecast in the model is the upper basin shortfall: the amount of water the upper basin would be unable to deliver from the Colorado River and would have to make up from its own sources. These shortages often amount to several million acre-feet per year, a large fraction of—and sometimes all of—the 4.5 to 5.4 MAF the upper basin is supposed to receive from the Colorado River between now and mid-century. Such shortfalls would cripple upstream farms and cities and devastate the economy of the upper basin.

Using different start years produces different results in detail, but the outcome is the same: with the bureau's assumptions, all is well on the river through mid-century and beyond. Using science-based assumptions, the Colorado River system always fails. Even using as a start year 1906, early in the wet period that preceded the signing of the compact, the system still fails by the mid-twenty-first century. Start the model in the Dust Bowl year of 1931 and failure comes much sooner.

—

On 13 December 2007, Interior secretary Dirk Kempthorne and reclamation bureau commissioner Robert Johnson signed a "record of decision" spelling out how the bureau would operate the Colorado River system between 2008 and 2026. Kempthorne pronounced the agreement the most important since the signing of the Colorado River Compact in 1922. At the end of the signing ceremony, held at Caesar's Palace in Las Vegas, the throng of water managers leaped to their feet and applauded. Under the plan, the bureau will coordinate operation of Lakes Powell and Mead, treating them as equally important. Nevada would pay to build a new reservoir in California to capture water in excess of the amount the United States is required to deliver to Mexico, further depleting the flow to the Colorado River delta and further endangering species there. The agreement encourages the basin states to augment the river by desalting seawater and transferring water from agriculture and from other river basins. In theory, cities such as Las Vegas and Phoenix could pay for desalting plants on the coast of California or Mexico and swap the new supply for a greater share of the Colorado River. Of course, California and Mexico would first have to agree.

The new shortage guidelines, resting as they do on the bureau's flawed computer modeling, are like a Potemkin village of water planning. The agency forecasts less than a 10 percent probability that, between now and mid-century, Lake Powell will fall below about 3,570 feet, 100 feet above the reservoir's power pool. But science-based modeling projects a much bleaker future. Time, and perhaps not much of it, will tell which prediction is right.

The existence of the agreement may be more important than its details. Mark Twain allegedly said that "whiskey's fer drinkin'; water's fer fightin' over." For almost the first time, instead of fighting, the basin states have voluntarily agreed to revise part of the Law of the River. As the reservoirs continue to fall, will this welcome comity continue? Once peace has broken out on the river, will it last? These questions I explore in chapter 18.

FIFTEEN

Rainmakers

AS IT BECOMES CLEAR THAT there is too little Colorado River water to support business as usual, governments and water agencies will pin their hopes on such remedies as desalting seawater, seeding clouds, building more reservoirs, and transferring in water from other basins. As long as there is the promise of augmenting or re-arranging supply, human nature will likely lead to a delay in doing anything about demand.

Desalting has an obvious appeal because 97 percent of the water on earth is seawater. Each year Nature desalts 35 billion acre-feet by evaporating water from the oceans and condensing it on land. That amounts to 5.5 acre-feet for each person on earth, four times the amount that hydrologists estimate humans require for all purposes, including the environment—but it is still not enough to prevent extreme shortages. The reason is that most fresh water is inaccessible: ice and snow lock up three-quarters of the earth's supply and underground aquifers hold most of the rest. Rivers and lakes, the only sources from which we can take water over the long term, account for only about 0.3 percent of earth's fresh water. But rivers and lakes are not distributed uniformly—Lake Baikal holds nearly one-quarter of the fresh water on the world's surface.[1] Even areas that have enough rainfall averaged over a year may receive most of it all at once: in only 100 hours, much of Asia gets nearly 90 percent of its annual monsoon rain.[2]

The United Nations estimates that by 2050 roughly 1.5 billion people will have less than 0.8 acre-feet of water per year, the level at which experts say true scarcity begins; the Yemenis and Palestinians have only about one-third that amount now.[3] Those who deny global warming or claim that it will have winners and losers (asserted only by those who expect to be among the winners), would do well to ponder a world in which 1.5 billion people have far too little water. Throughout history, wars have been fought and regimes have been toppled over water. A plan by Israel's Arab neighbors to divert water helped to trigger the Six Day War of 1967. Early in the twenty-first century, water is one of the core causes of the war in Darfur. But water scarcity could bring worse to that part of Africa. The Nile flows through ten different African nations, yet Egypt and Sudan have rights to all of it. Both countries have threatened to use force against any upstream nation that diverts the Nile for its own use.[4] Water is life and if there is no other choice, people will risk theirs to get it.

World population took until 1960 to reach three billion; to add the next three billion took only thirty-nine years. The world's population today is approaching seven billion and rises by 80 million a year. The problem for planners is that global water consumption has doubled every twenty years, twice the rate of increase in global population.[5] In 2006, the United States had 300 million people and by 2050 is expected to have 400 million. Population in the upper Colorado River basin is expected to expand from 9.5 million to 12 million by 2025, while the lower basin states, including giant California, grow from 42 million to 58 million.

Can desalting supply an appreciable amount of the water these extra tens of millions will need? If it is to do so, the cost of desalting will have to come down and several other inherent problems will have to be solved. Even then, how will desalting make a difference for cities located far from salt water, such as Denver, Phoenix, and Tucson? So far desalting has provided no more than a drop in the bucket of world water. More than 15,000 desalting plants in 125 countries produce 26 MAF of fresh water annually—but the world consumes that much in a few hours. If all the plants in the United States were up and running, they could desalt about 1.8 MAF per year—but that is less than 0.5 percent of total United States consumption.[6] California has proposals to build some twenty new plants; all those planned for Southern California running at full capacity could supply about one-half of Los Angeles's annual water needs.

Whether desalting can be scaled up enough to produce significant amounts of water depends in part on its cost. So far, desalted water has been too expensive for most countries, but the cost is dropping and today desalting is only about twice as expensive as capturing surface water and groundwater. Even if costs do not drop much further, desalting will look more competitive as water from other sources grows more expensive.

Most desalting plants use reverse osmosis, in which high pressure forces salt water through a membrane that blocks the salt, leaving fresh water on one side and brine on the other. The method requires pressures of around 900 pounds per square inch, 150 times as much as our home water systems.[7] To maintain that much pressure requires prodigious amounts of energy, thus explaining the cost of desalting. But the method has other inherent problems that will have to be overcome.

First, in many areas, America's shorelines tend already to be protected or if not, to be heavily developed. Finding an unprotected, undeveloped site in Southern California, for example, is hard enough; finding one where construction and operation of a desalting plant would not harm dunes, waterfowl, marine organisms, and the like is much harder. Even when a potential site can be found, some twenty-two different federal and state agencies must approve, not to mention various city and county departments.[8] This may lead some southwestern states to locate desalting plants in Mexico and swap the processed water for some of Mexico's share of the Colorado River—if Mexico would agree. Some desalting plants are apt to wind up offshore, like the oil drilling platforms off the coasts of California and the Gulf states.

The second problem is disposal of the brine stream emerging from a desalting plant. The only feasible dump sites are the ocean and deep underground. But unless brine dumped in the ocean is carried well out to sea, it may damage near-shore environments. Desalting also uses chemicals that corrode plant equipment and can enter the freshwater stream and render it unpotable.[9]

The third problem is of a different kind. As global warming reduces runoff, reservoir levels will drop and dams will produce less hydropower. More of the energy to fuel desalting plants would then have to come from burning coal and gas—and in the long run, from nuclear energy. Unless, that is, wind power can drive desalting plants, as is being tried at a new plant in Perth, Australia. In the short run, burning more fossil fuel to power desalting plants will increase greenhouse gas emissions, setting up a vicious cycle that will cause global temperatures to rise, reducing runoff and increasing the need for more desalting plants, and so on.

To understand what desalting is up against, consider the three largest desalting plants in the United States. The plant at Yuma, Arizona, took the Bureau of Reclamation seventeen years to build. After opening in 1992, the plant ran for six months at about one-third capacity and processed 23,000 acre-feet of water. After floods on the Gila River destroyed some of the canals that carried saline Wellton-Mohawk drainage water to the plant, the bureau shut it down. In the spring of 2007, a three-month run tested the plant's effectiveness. Running at 10 percent capacity for ninety days, the plant produced about 4,200 acre-feet of desalted water.[10] If it returns to operation, desalted water would be blended with Colorado River water above Morelos Dam and would count as part of the obligation of the United States to Mexico, requiring less water to be drawn from Lake Mead.

Because Santa Barbara, California, lies in a narrow swath between the Pacific Ocean and the San Rafael Mountains, it has neither the rain nor the room for large rivers. Most of Santa Barbara's water must come from rainfall—which averages only eighteen inches a year—and from groundwater. In the 1970s and again in the 1980s, severe droughts struck the central California coast. In spite of conservation measures that had cut water use by 40 percent, Santa Barbara's small reservoirs began to dry up, prompting voters to approve two relief measures. One was a pipeline to connect the city to the proposed central coast branch of the California State Water Project.[11] The other was a $34 million desalting plant that would produce 7,500 acre-feet a year, bringing the construction cost to $4,500 per acre-foot. No sooner was the plant completed than heavy El Niño rains filled reservoirs and replenished groundwater supplies. The cost of the desalted water proved so much higher than conventional sources that officials put the plant on standby. Additional conservation from low-flow toilets and low-water-use gardens further reduced consumption in Santa Barbara, eliminating the need for the costly desalted water. The plant was closed and some of its works sold and transferred to Saudi Arabia.[12]

In a 1955 article in *The Economist*, C. Northcote Parkinson promulgated his famous law: "Work expands so as to fill the time available for its completion." He found corroboration at the British Colonial Office, whose number of employees rose as the number of overseas colonies fell. Parkinson's Law has application in every field of human endeavor. One completely confirmed corollary is that "demand expands to consume supply."

Tampa, Florida, receives fifty-four inches of rain a year in a state with 50,000 miles of rivers and streams, 8,000 lakes and ponds, and 600 springs. The Florida Aquifer holds six billion acre-feet, twice the amount of the famed Ogallala Aquifer of the High Plains. Why would Tampa need a desalting plant? Ask Parkinson.

The Tampa Bay area, on the doorstep of Disneyland, is a magnet for new homebuyers, snowbirds escaping the icy northern winter, and tourists. By the mid-1990s, residents and visitors consumed over two hundred million gallons—615 acre-feet—each day. Groundwater withdrawals "depleted rivers, shrank ponds, drained swamps and wetlands, felled trees and caused saltwater intrusion into aquifers."[13] Projections showed that by 2030, the Tampa Bay area will add another one million people, all of whom will expect water to appear when they turn the tap.

In return for a state and federal subsidy of $270 million, the Tampa water authority agreed to reduce its groundwater pumping. To make up the missing water, officials elected to build a desalting plant projected to produce twenty-five million gallons of potable water a day, making it the largest municipal desalting plant in the Western Hemisphere. In one clever design feature, the plant would treat seawater that had already been used by an adjacent power plant. The proximity of the two plants would reduce the cost of transmitting the power needed to run the desalter.

The Tampa desalting plant opened in March 2003, leading optimists to begin to discuss opening a second in 2008. Then in May, the plant failed an important performance test. In February 2004, after the plant had produced three billion gallons—9,200 acre-feet—concern over design flaws and costs led officials to shut it down.

One reason for the higher costs was that exotic Asian mussels had clogged the plant's filters. The mussels reduced the expected lifetime—and therefore increased the replacement cost—of the membranes by about one-third. The use of chemicals to clean the fouled membranes also caused the plant to violate its sewer discharge permit.[14] In 2005, water officials announced that when the plant returned to operation, it would produce only fifteen million gallons per day. Tampa Bay Water had begun to reduce groundwater pumping in anticipation of the new supply; now it appeared the agency might have to increase groundwater withdrawals.

The plant was supposed to demonstrate the advantages of private ownership, but as the project stalled, a number of contractors went bankrupt and the Tampa Bay Water utility had to buy the plant. In November 2004, Tampa Bay Water hired a contractor to get the plant running again.

The water authority estimated the cost to restart the plant at $14 million, but the low bid, from a German-Spanish consortium called American Water, came in at $29 million.[15] Tests soon revealed that not only were mussels continuing to foul the membranes, acidic water was corroding the pumps. American Water first said that it would complete the repairs in October 2006. Then the date shifted to late December 2006, then to "sometime after January 1, 2007," then to no sooner than March 2007.[16] In April 2007 the plant again began production, its operators hoping it would reach capacity of twenty-five million gallons a day. A $48 million suit that Tampa Bay filed against the manufacturer of the membranes was settled out of court.[17]

The Pacific Institute sums up the lessons of Tampa Bay:

> The project's cost claims should caution desalination advocates against excessive optimism on price, and indeed, cost cutting is in part responsible for the project's difficulties. Moreover, the project had a number of unique conditions that may be difficult to reproduce elsewhere. For example, energy costs in the region are very low . . . compared to other coastal urban areas. The physical design of the plant . . . permitted the power plant to provide infrastructure, supporting operations, and maintenance functions. Salinity of the source water from Tampa Bay is substantially lower than typical seawater. . . . In addition, financing was to be spread out over 30 years, and the interest rate was only 5.2 percent.[18]

Desalting has another enemy: us.[19] Parkinson's Law could tell us that a new supply of desalted water will not relieve the stress on aquifers and existing surface water systems. Instead, as when a new freeway lane is added and quickly clogs with traffic, demand for water most likely will rise to consume the newly desalted supply, leaving us no better off than when we started. What will we do then—build another desalting plant?[20]

We may be able to solve, or if we cannot, to decide to live with, each of the problems associated with desalting. But one fact cannot be avoided: geography. How will a city like Phoenix, located far from seawater, benefit from desalting? Some cities may be able to desalt saline groundwater, but such sources are limited and will not last forever. In theory, a desert city might pay for a desalting plant on Mexico's Sea of Cortés and swap the water the plant produced for the same amount from Mexico's share of the Colorado. This would require that the cities have enough Colorado River water left to swap and that Mexico would agree. That would require a degree of international comity not seen when Colorado River water was in surplus.

Will it appear in the face of a prolonged water shortage? If not, the solution to the water crisis will have to come from some other source than desalting. Many western states persist in the belief that sprinkling clouds with chemicals will provide the missing water.

—

Attempts to persuade the gods to part with some of the rain they command surely goes back to the dawn of human society. Plutarch, in the first century A.D., observed that warfare appeared to cause rain. Nearly two thousand years later, Napoleon modernized Plutarch's theory, attributing the perceived increase in rainfall during battles to the concussion of cannon fire. Civil War troops, lugging artillery pieces across battlefield quagmires, agreed with the great general. The appearance of wet weather in the Great Plains just as the sodbusters arrived led to a rival theory, "rain follows the plow." Others thought that planting trees would bring rain; some state governments provided tax breaks for those who would do so. Disproving these loony notions had a great cost in broken dreams and human misery.

To put the concussion theory into practice, towns and farmers employed rainmakers to oversee the dangerous blasts. In the 1890s, Congress hired R. G. Dyrenforth to set off explosions to relieve parched west Texas. The inaptly named expert flew bags of dynamite aboard balloons and exploded them at altitude. His lack of success led to the nickname "Dry-Henceforth."[21] Like the patent medicine purveyors, snake-oil salesmen, and music men who roamed the gullible West, plenty of rainmakers were willing to take your money. By time the experiment concluded, they had moved on down the road to the next town and the next bunch of thirsty suckers. Where water is concerned, people will believe almost anything— and pay for it.

In 1836, meteorologist James Pollard Espy, aka "the Storm King," invented the theory of rain generation that we accept today. Espy's convection theory held that as air warms near the surface, it absorbs moisture; as the air rises it cools, condensing the water vapor as rain or snow. He tried to persuade Congress to fund a six-hundred-mile-long line of forest fires spaced twenty miles apart, claiming that the smoke would bring rain and confirm his theory. The plan had a built-in fail-safe mechanism: if it worked, the rain would put out the fires.[22] Wisely, the lawmakers declined to fund Espy's smoky experiment.

In 1915, the City of San Diego hired the self-proclaimed "Moisture Accelerator" and sewing-machine salesman Charles M. Hatfield. For a fee of $10,000, he guaranteed to create enough rain to fill the city reservoir. Hatfield built a tower that sent smoke from a secret concoction of chemicals and gases into the heavens above San Diego. Lo and behold, rain began to fall! And kept on falling—for five days. The target reservoir filled and overflowed; dams collapsed; bridges washed out, requiring sea launches to rescue the stranded passengers; trains derailed; houses left their foundations and floated away. Hatfield could not be reached to turn off the deluge—rain had downed the telephone lines. When he and his brother returned to San Diego and saw what they had wrought, they began to call themselves the Benson brothers—just in case. The city claimed that the deluge was an act of God and refused to pay "until it is determined that this is the direct result of Mr. Hatfield's efforts." The rainmaker pursued his fee all the way to 1938, when the court threw out the case.[23]

In the 1940s, Nobel laureate Irving Langmuir of the General Electric Company hired Vincent Schaefer as his research assistant, family financial problems having forced the young man to drop out of high school. One of Schaefer's assignments was to study whether dumping dry ice—frozen carbon dioxide—into clouds could produce snow. It seemed to, and soon atmospheric scientist Bernard Vonnegut, brother of late author Kurt Vonnegut, discovered that silver iodide crystals were more effective. Evidently, ice nucleated around the crystals and fell out of clouds as rain or snow. In spite of the promising beginning, according to a 2003 review by the National Academy of Sciences, more than sixty years of attempts to produce rain by seeding clouds with silver iodide have left "no convincing scientific proof of the efficacy of intentional weather modification efforts."[24] The fundamental problem is that rainmaking does not permit controlled experiments or replication, two essential features of scientific research. When rain does follow cloud seeding, who can say that it might not have rained anyway? Was the San Diego deluge of 1915 an act of God, or of the Moisture Accelerator? We have our suspicions, but no proof.

On the other hand, proving that cloud seeding does *not* work is just as difficult, guaranteeing eternal optimism and perpetual employment for rainmakers. A study by North American Weather Consultants, a member of the Weather Modification Association, reports that cloud seeding in the Rocky Mountains can produce up to 1.4 MAF per year—half the volume of the Central Arizona Project—for only $5 per acre-foot.[25] Wyoming is devoting nearly $9 million to its own seeding experiment; North Dakota is also trying the method. But Texas and Oklahoma found that seeding caused only 5 percent more rain to fall, leading them to conclude that teaching conservation would be the better investment.[26]

Even if cloud seeding does bring rain, no one can be sure how much will fall or where. In 1972, in Rapid City, South Dakota, as much rain fell in a few hours as normally falls in a year, raising floods that cost 238 lives. Earlier the same day, airplanes had seeded nearby clouds. A rainmaker said the seeding was not responsible because it was done "on the plains, 20 or 30 miles from the area where the heavy rains fell. . . . The flood would have happened anyway."[27] Who can say? Even if we could re-run the experiment, we would not.

Since seeding does not *create* rain clouds, but only taps moisture already present, if seeding does cause more rain to fall in one area, less may fall in another. Worried that rain may compromise the 2008 Beijing Olympics, China hopes to induce precipitation somewhere other than its capital. Over the last fifteen years, China has set up weather-modification offices around the vast country and launched weather satellites. Hearkening back to Napoleon, batteries of anti-aircraft guns will fire shells containing silver iodide at clouds approaching Beijing. What effect the crystals may have on the smoggy Olympic city we will find out.

Experts are skeptical. Dr. Robert Safarin of the National Center for Atmospheric Research worries about the side effects: "What if cloud-seeding isn't increasing snowpack, but is reducing it?" he asks. For Michael Cohen of the Pacific Institute, cloud seeding "is a giant waste of money."[28] The Bureau of Reclamation's nearly blank, one-page Weather Modification Web site reads, "This site is still under review by Reclamation senior management. It appears unlikely that it will be reposted."[29]

People want to live where it is warm and dry, where cities like Las Vegas, Los Angeles, and Phoenix can boast of having over well over three hundred days of sunshine a year and scores—even hundreds—of golf courses. The incongruity between where we want to live and where Nature supplies the water to sustain our lives is the root fact of existence in the West, the one simple fact of life in the West that is more important than any.

If desalting and cloud seeding seem unlikely to produce enough water to offset a diminished supply, the automatic response could be, "Build more reservoirs." This cry is already being heard around the West.

The least expensive way to increase storage is to heighten an existing dam, as has been proposed for example at Shasta Dam in Northern California. Raising its height only a few feet would allow the dam to impound several hundred thousand acre-feet of additional water. But raising the height of a dam presumes there will be surplus water to fill the additional volume. If not, the expense of adding storage will be wasted. Every so often, western-ers need to remind themselves of Luna Leopold's warning: "It would be physically impossible to add, on the average, more water to storage than the average annual flow of the stream."

Will California rivers have enough surplus water to fill an enlarged Shasta Reservoir and the other new dams that its governor has proposed? Snowpacks in the Sierra Nevada headwaters of the state's two major rivers, the Sacramento and San Joaquin, have already shrunk and started to melt earlier. By 2050, scientists project, average temperatures in the Sierras will rise by 2.1°F—twice the increase of the twentieth century. One group of scientists predicts that "in the Central Valley of California, it will be impossible to meet current water system performance levels; impacts will be felt in reduced reliability of water supply deliveries, hydropower pro-duction and in-stream flows."[30]

If California and the Southwest cannot make up their water deficits from their own streams and reservoirs, they may look to other basins nearby. They might turn their eyes north—to the Columbia, the Snake, and to the rivers that drain the western slopes of the Cascade Range in Oregon and Washington. But there they will find little relief. Since 1950, snowpacks in the Pacific Northwest have declined by as much as 70 percent.[31] The size of the reductions varies with elevation as they would if rising temperatures were the main cause. A study of the Columbia River basin predicts that we will have to choose between salmon and hydropower—there will not be enough water for both.[32]

Global warming will affect the Northwest even more than it does the Rockies. The Cascades are not a true mountain range, but a series of some-what isolated volcanic peaks. Cascade snowfields, at a lower elevation and moderated by the nearby Pacific Ocean, are already nearer to their melting point than Rocky Mountain snowpacks. If temperatures in the Northwest rise to the mid-range of the climate model projections, "You would have essentially no snow left in Oregon by April 1," says one scientist.[33]

If they cannot find relief in the Pacific Northwest, California and the Southwest may be tempted to look still farther north for water: to British Columbia and Alaska, where some of the largest rivers in the world run unchecked to the sea. There they will find waiting a fantastic water diversion and power-generating scheme, an apotheosis of reclamation known as the North American Water and Power Alliance. Though it has long been laid on the shelf, who can say that when the twenty-first-century drought becomes severe enough, NAWAPA and other gargantuan schemes may not be dusted off and brought up to date?

On a map of NAWAPA, giant rivers and canals course south from Alaska over the rest of the continent like serpents from the head of Medusa. Titanic dams on the Tanana, the Yukon, and the Mackenzie store vast volumes, then send it south to join water pumped up and over the mountains of British Columbia. The streams flow into a five-hundred-mile-long, 400-MAF reservoir in the Rocky Mountain Trench. As Glen Canyon disappeared underneath Lake Powell, so the mighty upper Columbia, Fraser, and Kootenai rivers would disappear beneath the monster reservoir. In addition, some of the arctic water would flow east into the Peace River, from there into the thirty-foot-deep Alberta–Great Lakes canal, and on 2,000 miles east to Lake Superior. By enlarging the canal, ocean-going vessels could travel from the Atlantic via the St. Lawrence and through the Great Lakes to central Alberta. The extra water would raise the elevation of the interconnected Great Lakes and spill over into the Illinois and Mississippi rivers. Cruise ships could sail from the Gulf of Mexico up to St. Louis and dock beneath Gateway Arch. But these dreams pale beside NAWAPA's plan for the West. Water from the giant Canadian reservoir would flow south and transmogrify the landscape of western North America, drowning the Great American Desert once and for all.

The water would rush down an eighty-foot-wide, fifty-mile-long tunnel drilled through Idaho's Sawtooth Range. Exiting at 30,000 cfs, the flood would speed on to California, the Southwest, and Mexico. To hold NAWAPA in one's mind, it helps to emulate Marc Reisner and suspend disbelief:

> Imagine a Pecos River Reservoir the size of Connecticut . . . another giant reservoir in Arizona. Imagine some 19 MAF of new irrigation water for Saskatchewan and Alberta. Imagine 2.3 MAF for Idaho, 11.7 MAF for the Texas high plains, 4.6 MAF for Montana, 13.9 MAF for California. . . . Imagine the Mojave Desert green. Mexico would get 20 MAF of water [and] triple its irrigated acreage.[34]

The gargantuan scheme had its inception in the 1950s, when a Los Angeles engineer brought his plan to Ralph M. Parsons, owner of a Pasadena construction company of the same name. Parsons fell in love with NAWAPA and established a tax-exempt, nonprofit foundation to promote the colossal project (though some said it was a tax dodge).[35] In the 1960s NAWAPA did not seem so fantastic. America had smashed the atom, defeated fascism, checkmated the Soviet Union, and was on its way to the moon. Why not rearrange the waterways of an entire continent?

A 1981 report put the cost of NAWAPA at $200 billion, equivalent to $500 billion in 2006 dollars.[36] Double that figure, or, more likely, recognize that estimating the cost of a project so vast and time-consuming is simply beyond our ability. A project on the continental scale of NAWAPA would certainly take until the mid-twenty-first century to complete and would certainly cost at least a trillion of today's dollars. On the other hand, the plan's hydropower dams would generate tens of thousands of megawatts *more* than required to operate its pumps and facilities. A spreadsheet model with a high enough discount rate and a long enough payback period could surely show that NAWAPA would earn a profit. Of course, the model would not include the cost to the environment—that would be incalculable. Luna Leopold described it this way: "The environmental damage that would be caused by that damned thing can't even be described. It could cause as much harm as all of the dam-building we have done in a hundred years."[37]

———

NAWAPA and similar giant-scale water transfers represent the ultimate in what Peter Gleick of the Pacific Institute calls the hard path to water security—the route we have followed up to now. On the hard path, demand drives supply. As population increases, more dams and infrastructure have to be built. But demand continues to rise, requiring still more infrastructure, and then more. Eventually, when there is no more supply, the hard path reaches a dead end.

The other path—the soft path—

> strives to improve the productivity of water use rather than seek endless sources of new supply. It delivers water services and qualities matched to users' needs, rather than just delivering quantities of water. It applies economic tools such as markets and pricing, but with the goal of

encouraging efficient use, equitable distribution of the resource, and sustainable system operation over time. And it includes local communities in decisions about water management, allocation, and use.[38]

There is no doubt that in the century of global warming, gaining water security in the West is going to cost a great deal of money. The question is whether we choose to spend that money once, or over and over. Ultimately, humans must learn to survive on the water that Nature supplies. Ultimately, only one path is sustainable—the soft path. We know how to follow the soft path and we have the technology—but do we have the will?

SIXTEEN

Let People in the Future
Worry about It

AS THE FLOW OF THE COLORADO RIVER declines and its reservoirs shrink, two inevitable flaws of irrigation will grow steadily worse. The less water in the river, the more silt and salt will compromise its use.

Start with silt. Since the time of the Conquistadors, El Río Colorado has been known for its mud. "Too thick to drink, too thin to plow" was the way later arrivals sometimes described the river. In 1916, E. C. LaRue reported that the Colorado at Yuma carried 160 million tons of silt a year, enough to fill 5,000 modern ship cargo containers—*each day.* Before the big dams, the Colorado carried that silt all the way to the Gulf of California, where it built one of the world's great ecological wonders: the Colorado River Delta.

It is hard for us to conceive just how much material the Colorado River has excavated, carried, and deposited. We can estimate how much rock the river has eroded from the Colorado Plateau, or we can study the other end of the process, where the Colorado ran out of energy and deposited silt to fill the Salton Trough, a buried geologic feature stretching from the Gulf of California to beyond the Salton Sea. Drilling for oil and gas has shown that the trough's sediment fill often extends downward for one or two miles. The surface area of the Salton Trough measures about 130 miles by 70; thus if the fill averages one mile thick, its volume calculates to thirty billion acre-feet—more than a thousand times the volume of Lake Powell.

Not only does the river carry a vast volume of silt, the amount varies inexplicably from year to year. From 1925 through 1940, the sediment gauge at Lee's Ferry showed an average of 195 million tons of silt per year, 20 percent more than LaRue's 1916 calculation. But from 1941 to 1957, the load dropped by more than half, to 86 million tons per year.[1] Scientists are not sure why, though there is evidence that flash floods, the main source of silt in the Colorado River, were more common before 1941.

Knowing how much sediment the Colorado River carries is important for several reasons. For one, the information allows us to estimate how long it will take Lake Powell to fill. For another, scientists need to monitor and understand sediment loads in order to determine whether species along the river have enough for critical habitat. In 1957, while construction was under way on Glen Canyon Dam, eighteen sediment gauges had been operating on the Colorado River for more than ten years—the longest and most thorough set of measurements of sediment transport anywhere in the world.[2] But once the gates at Glen Canyon closed, the Department of the Interior lost interest in tracking sediment loads and the Geological Survey began to shut down the gauges one by one. The Grand Canyon gauge closed in 1972, and by 1989, only two of the long-running sediment gauges were still operating. By then, it had become apparent that the loss of sediment in the Grand Canyon was endangering its wildlife, but the near-absence of gauges and the lack of longitudinal data made a remedy difficult to identify.

In 1986, the bureau turned to sonar to measure the amount of silt on the floor of Lake Powell.[3] The study found that 868,000 acre-feet had accumulated, most of it in a delta at the head of the lake, but some in a small wedge piled against the upstream face of the dam. Over the lifetime of the lake, the average accumulation calculates to 37,000 acre-feet per year (about 45,000,000 million tons). That would give Lake Powell about seven hundred years before it fills completely. Boosters of the lake have made much of this comfortably distant date, but the shrunken reservoir of the future will render the number meaningless.

Were Lake Powell to remain half full, the same input of silt would fill it twice as fast. But as shown in chapter 14, increasing demand and shrinking supply are going to deplete Lake Powell. At dead pool, it will hold 1.9 MAF of water. To fill that volume at 37,000 acre-feet per year would take not hundreds of years, but fifty-one years. Were the higher loads of the 1920s and 1930s to return, the lake would fill even faster. But another, recently discovered effect may further shorten Lake Powell's life span.

Rafting down the Colorado River through Cataract Canyon today, one sees the banks gradually turn from bedrock to thin layers of loose sediment. Nearing Lake Powell, the sediment banks seem to grow higher as the river cuts its way down through the soft strata. These thin sediment layers are the delta of Lake Powell, now exposed and eroding due to the reservoir's lower elevation. Two hundred miles downstream in the Grand Canyon, approaching Lake Mead below Separation Canyon, one sees the same silt layers. The deposits are decades old at most and so loose that they crumble and fall under their own weight. As you watch, hundreds of tiny rivulets of sand—miniature silt-falls—trickle down from the tops and sides of the banks. Less often, a chunk as large as a car slumps down in a cloud of dust. The next flash flood washes the loose silt into the river, which carries it down to the lakes and uses it to rebuild their underwater deltas. Thus a smaller Lake Powell receives not only the new silt that the river continues to wash down to it—but also the recycled, older silt from the exposed delta deposits. The effect reduces Lake Powell's projected lifetime still further, though by how much no one can be sure.

The surface of a delta slopes gently downward near shore, then steepens toward the deeper water. Silt can rush down the steep delta face and flow along the bottom for scores of miles. Some of the silt that the sonar study found wedged against the face of Glen Canyon Dam had traveled the entire length of the lake. Some also came from farther down-lake, from the Dirty Devil, Escalante, and San Juan tributaries.

As the silt at the dam face continues to rise while the surface of the lake falls, the reservoir becomes squeezed from both directions, possibly leading to bizarre outcomes uncontemplated by dam-builders. Imagine a time when the surface of Lake Powell has fallen below the dam power intakes at 3,470 feet, so that no water can exit through them. Imagine that silt built up on the lake bottom and against the dam face blocks the river outlet tubes at 3,374 feet. Now the water in the remnant of Lake Powell has no way out. Except for the trickle from the Paria River, between Lee's Ferry and the mouth of the Little Colorado, the mainstem Colorado River in the Grand Canyon would run dry.

In preparing *Cadillac Desert,* author Marc Reisner paid a call on the bureau's Office of Sedimentation. The writer asked what the agency planned to do to counteract the inevitable buildup of silt. "All of our bigger reservoirs were built with a sedimentation allowance," his bureau host replied. "There's enough surplus capacity in them to permit most of the projects to operate according to plan over their payout lifetime. In most cases that's fifty to one hundred years. After that, silt will begin to cut into capacity. It hasn't happened yet to any significant degree." But, Reisner wanted to know, what is the bureau's plan when the reservoirs do fill? "We're working on it," came the response.[4]

Whatever the Office of Sedimentation had in mind, there are only a few ways to prevent a reservoir from silting up and none of them will work at Lake Powell. One method is to flush muddy water through openings in the dam. But that would foul the generators and be unavailable in any case once the lake level has fallen below the generator intakes. Some silt could be flushed through the dam's outlet works but not enough to have any effect on silt entering 186 miles upriver. A second solution is to build another dam upstream to trap incoming sediment, but that is how we got here and would eventually produce two filled reservoirs instead of one.

The most obvious strategy is to dredge the sediment, haul it away, and dump it. Suppose that society decides to live with the sediment already in Lake Powell but sets itself the task of preventing more from accumulating. How big a job is that?[5] Assume that Lake Powell receives 45,000,000 tons of silt each year, the approximate equivalent of the 37,000 acre-feet found in the sonar studies. To be efficient, one would want the largest dump trucks money could buy, models that might cost $400,000 and hold forty tons.[6] The trucks would pick up their loads at the dredgers, travel slowly to the dump site, unload the sediment, and return for the next load. Within range of Lake Powell there is only one site whose host state or nation might be willing to accept that much sediment—and might actually benefit—Mexico's Colorado River Delta, roughly 650 miles away on the Gulf of California. If the trucks could manage one round trip every two days, to keep up with the 120,000 tons of silt that the river delivers *each day* would require 6,000 of the behemoths, bringing the total purchase price to $2.4 billion. To keep each truck operating might cost $1 per mile, or $1.4 billion a year. Replacement costs would be in the scores of millions each year. The project would require roads, dredging and ancillary equipment, housing for workers and drivers, and so forth. Thousands of tanker trucks would have to haul in the diesel fuel to slake the thirsty monsters. Even if the money could be found, dredging on such a scale would destroy Lake Powell in order to save it.

Thousands of giant trucks rolling out of Glen Canyon each day would obliterate the local environment and everything on either side of the road to the dump site. And since the Colorado River never rests, neither could the dredgers and the giant trucks. Sisyphus would have had it easy in comparison.

Another strategy would be to pipe slurried silt two hundred miles from the head of the lake around the dam and dump it into the Colorado River below Glen Canyon Dam. Here, at least, gravity would be an ally. Were the job defined as preventing the further buildup of silt in Lake Powell, the pipelines would have to move thirty million gallons of slurry each day, a huge amount. And again, since the river never rests, neither could they.

Aldo Leopold foresaw the end result: "The rivers on which we have built storage reservoirs or power dams deposit their deltas not only in the sea, but behind the dams. We build these to store water, and mortgage our irrigated valleys and our industries to pay for them, but every year they store a little less water and a little more mud. Reclamation, which should be for all time, thus becomes in part the source of a merely temporary prosperity."[7] Like the cholesterol in our veins, silt steadily builds up until it clogs a reservoir completely. The strategies above show that removing the silt in western reservoirs, even were it physically possible, would be incalculably expensive. Instead, with the possible exception of some of the smaller reservoirs, those across the West are destined to fill with silt and nothing can be done about it. Instead of blue lakes providing power, water, and recreation, one day hundreds of mud-filled cavities will pockmark the West.

To convince us of the inevitability of this future we need an example of a reservoir that has already filled with silt. Matilija Dam and Reservoir in Ventura County, California, will serve well. They reveal the future of every dam and reservoir, as though on ultra fast-forward.

Matilija Creek rises in the mountains west of Ojai and joins other mountain streams to make the Ventura River. Only sixteen miles below the dam, just west of the city of Ventura, the river enters the Pacific. Even though the drainage area of Matilija Creek comprises only about fifty-five square miles—35,000 acres—the pre-dam river was a major source of sediment for beaches along the coast to the south, toward Los Angeles, as well as a prime natal stream for ocean-running steelhead.

Figure 24. Matilija Dam in full flood, illustrating the fate of dams (Courtesy of Paul Jenkin, Matilija Coalition)

Having gotten the reclamation religion, in the 1930s Ventura County officials began to yearn for a dam on Matilija Creek to provide flood protection and water for agriculture. In 1941, the Army Corps of Engineers reported that the projected cost of the dam would outweigh its benefits. When, in spite of the report, county supervisors placed a bond measure on the 1945 ballot, voters approved. The project was to cost $682,000. Construction began in 1946 and less than two years later, Matilija Dam, nearly 200 feet high and 620 feet wide, was in place. When the cost turned out to be $4 million, Ventura County sued the design engineers but lost. The reservoir's fundamental problem was geometric and thus known from the start: the dam needed to be tall and therefore would be costly, but because of the upstream topography the reservoir could hold only 7,000 acre-feet of water. Silt from Matilija Creek would not take long to fill such a small volume. No sooner was the dam completed than a forecast gave it a life expectancy of thirty-nine years.[8]

In 1964, Bechtel Corporation condemned the dam and recommended removing it, estimating the cost at $300,000. Instead, to relieve the pressure of thousands of tons of silt, Ventura County elected to cut a notch in the dam crest 285 feet wide and 30 feet deep. In 1978, the notch had to be enlarged to 358 feet, more than half the width of the dam. The notch allowed water to leave at a lower elevation, but reduced the capacity of the reservoir and sped up the time until sediment filled it. By 2000, silt occupied 93 percent of Matilija Reservoir. In the winter rains, water wanders across the top of the dam and cascades down its face. Grass, weeds, and small trees grow out of cracks, revealing the ugly visage of dams in old age.

In sixty years, Matilija has gone from essential to useless to dangerous. Today, its problems include the following:

> Large volumes of sediment deposited behind the dam and the loss of
> the majority of the water supply function and designed flood control
> capability; the deteriorating condition of the dam; the non-functional
> fish ladder and overall obstruction to migratory fishes; the loss of
> riparian and wildlife corridors between the Ventura River and Matilija
> Creek; and the loss of sediment transport contributions from upstream
> of the dam.[9]

In places the main coastal artery between San Francisco and Los Angeles, Highway 101, passes perilously close to ocean beaches. As the last phrase in the preceding extract indicates, without the sand that Matilija Creek formerly provided, these beaches are eroding toward the highway. Surfer's Beach, down the coast from the mouth of the Ventura River, has turned to cobble. As beaches and dunes shrink, habitat for species, some of them endangered, also shrinks.

In the late 1990s, a coalition of government agencies and environmentalists developed a plan for removing Matilija Dam.[10] The estimated cost of the removal is $128 million, four times in constant dollars what it cost to build the dam. About one-third of the sediment fill will be slurried and piped around the dam. The remaining two-thirds will be recontoured to allow storms to wash it gradually and harmlessly downstream. A new fish passage through the sediment pile will provide steelhead with a route to their birthright headwaters. Then engineers will dynamite Matilija Dam and remove it in fifteen-foot sections. It will be the largest dam ever dismantled.

Lakes Powell and Mead each hold roughly 10,000 times the volume of the Matilija Reservoir. Though scaling up by four orders of magnitude borders on the absurd, *if* we did it, to fix one of these two reservoirs would cost $1.3 trillion, 10 percent of Gross Domestic Product of the United States.

Reservoir silt can contain fecal matter and metals such as arsenic, boron, lead, mercury, and selenium, some of it natural and some accumulated from decades of irrigation runoff. The lower Lake Powell falls, the more likelihood that water exiting through the outlet works will entrain some of this noxious mess and send it down the Grand Canyon.

There is another reason why the best place to view the Grand Canyon in the future might be from the safety of the rim. Engineers design concrete arch dams to hold back the weight of millions of acre-feet of water. But the same volume of wet sediment weighs roughly twice as much and transmits stresses differently. The amount of silt that could fill the space behind Glen Canyon Dam would weigh roughly 70 billion tons. Will the dam hold that much mud? No one knows.

When asked in 1995 what the bureau intended to do when Lake Powell filled with sediment, Floyd Dominy laughed and said, "We will let people in the future worry about it."[11] To paraphrase Pogo, we have met them and they are us. Because the bureau built most western dams within the same fifty-year period, many are going to silt up within the same time span.[12] On the Colorado River, Dominy's people in the future will find sixty million acre-feet of dangerous mud trapped behind concrete arches designed to hold water, not mud, and find they can do absolutely nothing about it.

—

Just as natural law requires that reservoirs fill with silt, so continued evaporation and reuse require that irrigation water becomes increasingly saline. A certain amount of salt—not only sodium chloride, but also calcium, potassium, magnesium, sulfate, and bicarbonate—is essential to life, partly for the trace elements the salts carry with them, but too much salt is poison.[13] Plants must expend vital energy excluding salt, leaving them less for growth. Crops stunt, forcing irrigators to switch from profitable winter vegetables and citrus fruit to salt-tolerant crops of lower value and productivity, especially alfalfa. Unless salt is removed, eventually it poisons the soil.

Through the ages, irrigators have typically flooded their fields to a depth of several inches. But three-fourths of the water evaporates from soil and plant leaves, leaving all the salt behind. Of the one-fourth that sinks into the soil, some water emerges downstream, where a farmer uses it to irrigate his field, after which it sinks into the ground again, to emerge farther downstream, and so on. By the time Colorado River water reaches Mexico, its has been reused several times, each time growing saltier—then Mexican irrigators use it again.

If the top of the zone of water saturation—the water table—lies within a few feet of the surface, capillary action draws groundwater upward and can waterlog the soil with saline groundwater. In many western irrigation districts, farmers must lay pipes below ground to drain off the salty effluent. Some carry the waste to evaporation ponds, where the selenium it contains can deform and kill wildlife, as at the infamous Kesterson "wildlife refuge" in California's Central Valley Westlands Water District.

Before the Bureau of Reclamation built Hoover and Glen Canyon dams, the salt content of the virgin Colorado River increased from 50 parts per million in the Rockies to 380 ppm at the border with Mexico. Much of the natural salt comes from erosion of unusual evaporite deposits in the Colorado River basin. After Hoover Dam allowed large-scale irrigation in the lower basin, the salinity of the Colorado River at Imperial Dam just above the Mexican border rose to 785 ppm. In 1944, the United States agreed to deliver 1.5 million acre-feet of Colorado River water annually at the border. Thus Mexico, once the destination of 100 percent of the water in the lower Colorado, had to settle for 10 percent.

Before the gates closed at Glen Canyon, the upper basin used relatively little water, leaving a surplus of several million acre-feet to cross the border. That was enough to sustain Mexican farmers and the Colorado River Delta. But as soon as Glen Canyon Dam came online, the priority of the United States became to fill Lake Powell as fast as the Law of the River would permit. The bureau immediately limited the delivery to Mexico to the statutory 1.5 MAF, ending the surpluses.

At about the same time, the Wellton-Mohawk irrigation project, one of the last in Arizona, began to add highly saline water to the lower Colorado River, raising the salt level at the border to 1,500 ppm. Mexico protested, only to be told that the 1944 agreement had said nothing about water quality, only that the delivery to Mexico "shall be made up of the waters of said river, whatever their origin."[14] The conflict led President Luis Echeverría to threaten to take the United States to the World Court. The risk to America's Mexican oil deliveries led President Nixon in 1973 to send Herbert Brownell, a former attorney general, south to negotiate. Within a few months, the two nations had agreed that the salinity of water at the border would be no more than 145 ppm higher than it was at Imperial Dam in 1976: 879 ppm.[15] Thus the salinity of water reaching Mexico is not to exceed 1,024 ppm, good enough for agriculture but not by much. The World Health Organization standard for drinking water is 500 ppm.

The salinity of the Colorado varies inversely with the amount of water the river carries, as though it has a fixed amount of salt and the lower the flow, the higher the concentration, and conversely. About half the salt comes from natural sources and most of the rest from agricultural runoff. The bureau estimates that the damages from salinity amount to about $300 million per year, another cost not considered in the original benefit-cost analysis for the CRSP, or for most other large dams.

As the flow of the river has declined during the twenty-first-century drought, salinity at Imperial Dam has risen to about 700 ppm. It would have risen higher and likely abrogated the Mexican treaty had the bureau and other federal agencies not introduced a number of salinity control measures. These include diverting and gathering seeps from saline springs, improving irrigation methods, and lining canals. Altogether, the bureau estimates that such methods have prevented about one million tons of salt from entering the Colorado River system annually.[16]

To forecast future increases in salinity, the Bureau of Reclamation uses the same computer model and the same assumptions as in its other studies of the Colorado River. Since salinity varies inversely with water volume, the more water the bureau assumes will be present in the river, the lower the projected salinity. Not surprisingly, the agency forecasts that salinity will continue to remain below the treaty limit.[17] But an independent study found that a decrease in runoff of only a few percentage points would drive salinity over the limit.[18] As the flow of the lower Colorado River declines, breach of the treaty limit appears inevitable.

◆

A river is a living scientific demonstration project. The laws of gravity and physics determine that water erodes, flows downhill, and when it runs out of energy, deposits the sediment it carries. The laws of chemistry ensure that water extracts salt from rock and soil and as the water evaporates, leaves the salt behind. These laws doom each dam to eventual failure. Meanwhile, in the much shorter run, for the native creatures that depend on a river for their lives, a dam makes everything worse.

SEVENTEEN

A Hundred Green Lagoons

THE COLORADO RIVER SPEEDS through the Grand Canyon fast enough to create dozens of boulder-filled, white-water rapids, but in the backwater eddies it slows and deposits silt for fish and wildlife habitat. Since 1963, when the gates closed at Glen Canyon, the giant dam has trapped virtually every grain brought down by the mainstem river. Whatever silt remains below the dam today for riverine habitat has to come from tributaries within the canyon—mainly the Paria and Little Colorado rivers—and they do not provide nearly enough. The lack of silt, the cold temperature of the water, its unnatural flow schedule—all are driving species in the Grand Canyon to extinction, in clear violation of the Endangered Species Act.

Before the dam, the temperature of the river varied from near freezing in the winter to eighty degrees in the dog days of late summer. Today, as any rafter can tell you, no matter the season, the water in the Grand Canyon is cold. Glen Canyon's generator intakes sit 230 feet below the nominal surface of the lake. At that depth, the water of Lake Powell and therefore the water emerging from the dam has a nearly constant temperature of 46°F. On its three-hundred-mile journey through the Grand Canyon to Lake Mead, the cold lake water so dominates the warmer but smaller tributary inflows that the river heats by only a few degrees. Native fish that evolved in warm, silty water often cannot reproduce in the cold, clear water. On the other hand, conditions are just right for nonnative species like rainbow trout, which consume the young of the native species. The stretch of the river below Glen Canyon has become a world-class rainbow fishery.

In the late spring and summer, discharge in the Grand Canyon once reached 100,000 cfs or more; in winter, the flow would fall to a few thousand cubic feet per second. During the spring snowmelt, the volume would swell rapidly in a few weeks and fall as quickly, stranding sediment near the high water mark and rebuilding habitat. The dam replaced the gradual seasonal variations with sharp daily and hourly ones and put the Colorado River in the Grand Canyon under the command of power managers, whose job is to sell maximum power at peak rates. Once the sun rises on the desert metropolises, appliances and air conditioners come on and electricity demand rapidly ramps up. Hydropower dam operators respond by turning a valve, allowing more water into the penstocks and out through the turbines, instantaneously producing just the amount of power needed. To generate the power increase would take a coal-fired plant several days.

During the 1970s and 1980s, releases from Glen Canyon Dam often varied by 12,000 cfs within twenty-four hours and sometimes by as much as 20,000 cfs. Within twenty minutes, flows would ramp up by thousands of cubic feet per second. Surging through the Grand Canyon, the high flows raised the river level as much as ten feet in a few hours. Rafters tying up for the night on a Grand Canyon sandbar had to know the release schedule, otherwise the rising water might float their raft loose and send it downstream, or falling water might strand it high on the bank. The rushing, silt-free surges scoured the channel and banks of the river, washing sediment downstream and out the mouth of the Grand Canyon. Vital sandbars shrank, taking with them the habitat needed to sustain the canyon's unique ecosystem. The denizens of the Grand Canyon confronted a cold, clear river that ebbed and flowed on a schedule set by human activity in distant cities.

A reading of the founding legislation of the Colorado River Storage Project makes the fealty to hydropower seem odd. After all, the act specifies that power generation is to be merely an "incident" of other purposes such as flood control and irrigation. But the act also instructed the secretary of the Interior "to produce the greatest practicable amount of power and energy that can be sold at firm power and energy rates."[1] Located at the foot of the upper basin, Glen Canyon Dam has no direct role in flood control or irrigation. As long as dam managers delivered 8.23 MAF of water each year, they could do so on whatever hourly, daily, and weekly schedule they wished.[2] It did not matter, for all the water wound up in Lake Mead. From there, it passed through Hoover Dam and on to downstream users.

The next major piece of legislation, the Colorado River Basin Project Act of 1968, suggested new priorities. Its purposes were as follows:

> Regulating the flow of the Colorado River; controlling floods; improving navigation; providing for the storage and delivery of the waters of the Colorado River for reclamation of lands, including supplemental water supplies, and for municipal, industrial, and other beneficial purposes; improving water quality; providing for basic public outdoor recreation facilities; *improving conditions for fish and wildlife,* and the generation and sale of electrical power as an incident of the foregoing purposes.[3]

The notion that a Bureau of Reclamation project should attempt to improve conditions for fish and wildlife was new. The obligation drew mention even before hydropower, which was again relegated to incidental status. A literal interpretation would have Glen Canyon Dam managed so as to accomplish goals other than hydropower first, including protecting fish and wildlife. But as Senator Paul Douglas pointed out during the Echo Park debate, the balance sheet for the entire CRSP depended on cash from Glen Canyon. The bureau ignored the directive to protect fish and wildlife and continued to release water solely to generate peak power, with its attendant hourly fluctuations and harm to the ecosystem of the Grand Canyon.

Filling Lake Powell proved to be such a logistical, political, and hydrological conundrum that dam managers had few options. Many people, including Senator Paul Douglas, doubted there was enough water to fill it. To do so, Glen Canyon Dam had to hold water back. Yet to obey the Law of the River, the dam had to release millions of acre-feet downstream. Would there be enough water to meet these contradictory requirements? The upper and lower basins had a vital but opposite interest in the answer.

Of course, until the reservoir gained 1.9 MAF and rose above the outlet works, the point was moot, for the dam could release no water. Until the reservoir gained another 2 MAF and topped the generator intakes, the dam could generate no power. In spite of the difficulty, to surmount the penstocks took only a little more than a year. To fill the lake entirely would require holding back another 20 MAF, water that could not be sent downstream and therefore could generate no cash. Only the secretary of the

Interior—then Stewart Udall—had the authority to resolve the dilemma. Thanks both to good management and good luck, Lake Powell did fill, though it took until 1980, just in time for the 1983 flood.

At least engineers could calculate how long it might take to fill the reservoir and decide how much water to release and when. In contrast, in the absence of biological studies, no one could say just what would "improve conditions" for fish and wildlife in the Grand Canyon. No one could demonstrate that the topsy-turvy operation of Glen Canyon Dam actually harmed species or if it did, what changes might undo the harm.[4]

———

Prior to the National Environmental Policy Act of 1969 (NEPA), federal agencies did not have to consider the effects of their programs and facilities on the environment.[5] Indeed, as the agencies pointed out, such studies were not authorized.[6] Though the bureau built Glen Canyon Dam before NEPA, conservationists in the 1980s concluded that the release fluctuations were harming fish and wildlife downstream. The Environmental Defense Fund (now known as Environmental Defense) and recreational users sued, arguing that since the bureau continually made decisions about dam operations, NEPA should apply. The government responded that the operation of the dam was merely business as usual and therefore the door to NEPA should remain shut, no matter the damage to the environment.[7] The court agreed; more lawsuits and citizen protests followed, but to no avail.

By the early 1980s Glen Canyon's generators were two decades old and their wire coils needed rewinding. President Ronald Reagan had mandated a study of existing hydropower dams with a view to increasing their capacity. That led his secretary of the Interior, James Watt, to propose adding new generators to the old river bypass tubes at Glen Canyon. When that idea failed to garner support, Watt suggested enlarging the existing generators.

Not every western politician loves wilderness camping, but those who do not are usually smart enough to keep quiet about it. Not Watt. After a rafting trip through the Grand Canyon, the Interior secretary said, "The first day was spectacular. . . . The second day started to get a little tedious, but the third day I wanted bigger motors to move that raft out. There is no way you could get me on an oar-powered raft on that river—I'll guarantee you that. On the fourth day we were praying for helicopters, and they came." Watt summed up his attitude toward the outdoors: "I don't like to paddle and I don't like to walk."[8]

The larger generators that Watt had in mind would lead to even greater release fluctuations and more negative downstream effects, likely triggering NEPA. The key expectation of NEPA is that an environmental impact statement (EIS) must precede any new project. But there is an alternative: if an "environmental assessment" shows "no significant impact," NEPA does not require a full-blown EIS. To bureau officials an assessment was much preferable, for an impact statement would make Glen Canyon the first large dam to have to account for its effect on the environment, a frightening precedent for any reclamation agency. An assessment also had the advantage of not requiring public input.

To provide scientific data for the assessment, Watt established a program called Glen Canyon Environmental Studies—but with conditions. Scientists were to explore only the impacts of dam operations on recreation, sediment transport, and biology in the Grand Canyon. The study was merely to report findings, not recommend changes in the operation of Glen Canyon Dam. Also off limits were aesthetic, cultural, and "non-use" considerations—that is, being able simply to gaze at the sublime vistas of the Grand Canyon was not a protected environmental activity.

To manage the science studies, officials of the Bureau of Reclamation appointed one of their staff biologists, a young man named David Wegner. His bosses said Wegner was "window dressing," a "token environmentalist," and that "the less he accomplished the better." One staffer told Wegner that the bureau had hired him rather than an engineer because "an engineer might get something done." Dan Beard, Bill Clinton's iconoclastic reclamation commissioner and a veritable anti-Dominy, later said, "They didn't even want [Wegner] to list [his] office phone number."[9] But Wegner had the courage of his convictions. Instead of confirming that the proposed generator upgrade would have "no significant impact," his studies revealed just the opposite. They showed that for the previous two decades, the operation of the dam had been harming recreation and wildlife. The larger, rewound generators would allow even greater release fluctuations and do even more harm.

Prior to the assessment report, scientists had assumed that dampening the extreme release fluctuations would reduce erosion in the Grand Canyon and help restore the sandbars. The environmental assessment showed that instead, even with steadier flows, the sand would stay in the channel, below water level, rather than on the bars and beaches.[10] The report recommended further studies but, as forbidden to do, did not propose changes in the dam release schedule.

The public continued to clamor to be heard. When the bureau's partner in operating Glen Canyon Dam, the Western Area Power Administration, proposed a new power-marketing plan, environmentalists again sued. This time, they won.[11] Don Hodel replaced Watt as secretary of the Interior; under President George H. W. Bush, Manuel Lujan succeeded Hodel. Meanwhile, the Department of the Interior commissioned the National Academy of Sciences to review the department's environmental studies of its Glen Canyon operations. The academy report concluded that dam operations were harming both the ecology of the Grand Canyon and its use for recreation. After the court required public hearings, Lujan decided to order a full-blown EIS. But with the damage ongoing, environmentalists and recreational users continued to press the secretary and Congress. In 1991, Lujan approved a new release schedule with reduced fluctuations. Conservation groups and several congressional leaders grew concerned that the bureau was dragging its feet in completing the EIS. In 1992, Congress passed the Grand Canyon Protection Act, which directs the secretary of the Interior to manage Glen Canyon Dam "under existing law in such a manner as to protect, mitigate adverse impacts to, and improve the values for which Grand Canyon National Park and Glen Canyon National Recreation Area were established, including, but not limited to natural and cultural resources and visitor use."[12] Congress had come a long way since the debate over Echo Park, when wildlife and ecosystems earned hardly a word of discussion. Now the environment of the Grand Canyon, on paper at least, was on an equal footing with power generation and water delivery. To complete the EIS took until 1995; a year later Interior secretary Bruce Babbitt signed a "record of decision" mandating a program of adaptive management under which the bureau would incorporate future research findings into its operation of the dam.

Thus by the middle of the 1990s, power generation at Glen Canyon had to compete with other important values: restoring critical sandbars and habitat in the Grand Canyon and protecting endangered species. But how could the bureau accomplish these conflicting goals? Once a high dam is in place, only one variable remains to be manipulated: the amount of water the dam releases.

In the warm, muddy waters of the virgin Colorado River, only highly specialized fish species, and not many of them, had been able to evolve.[13]

Figure 25. The humpback chub (U.S. Geological Survey)

In John Wesley Powell's day, from snow to sea the Colorado River held only thirty-five fish species; an eastern stream of similar size might have had as many as two hundred. The native fish are mostly minnows, chubs, and suckers, but their unique habitat has made them unlike any members of those families we are likely to have seen. The Colorado pikeminnow is over three feet long and lives for fifty years. It is now listed as endangered in the Colorado River overall and no longer swims in the Grand Canyon. The Quasimodo-like chubs range up to twenty inches, with large dorsal humps and long tails. Their size and front-loaded shape are thought to help them hold their position in the swift waters.

The mainstem Colorado River in the Grand Canyon originally supported eight native fish species; today four remain.[14] The humpback chub is protected under the Endangered Species Act of 1973, but the bureau's Glen Canyon Environmental Studies report showed that the dam had sent the chub into serious decline. Later, the bonytail chub and the razorback sucker, the latter once so plentiful that farmers ground up the fish for fertilizer, were also listed.

Reducing the hourly release fluctuations turned out to make no difference to the ecology of the Grand Canyon. So much sand had been lost that the bars needed not only to be rearranged but also augmented with new sand. To figure out how to provide it required detailed knowledge of the historic sediment budget throughout the Colorado River and its tributaries—but only two of the long-running gauges were still operating, forcing scientists to rely on educated guesses.

The Paria and Little Colorado rivers are the only significant sources of new silt in the Grand Canyon. Before Glen Canyon Dam, the two (plus other minor sources) provided 16 percent of the sediment that traveled through the canyon.[15] As noted, with the dam having prevented high flows, the sediment contributed by the two tributaries tended to remain lower down in the river channel, below the surface. The scientists reasoned that if Glen Canyon Dam were to release its maximum flow just after the tributaries had deposited sediment into the mainstem channel, the high water might lift the silt onto the sandbars and, when the water receded, strand it there.

In 1996, operators opened Glen Canyon's outlet works and generators to their maximum. Like a fire hose of the Titans, 45,000 cfs arced scores of feet up and out before falling back to the river below. During the eight days of high flows, scientists kept their fingers crossed. When the releases returned to normal, would the sandbars have gained sediment? Sadly, the answer turned out to be no.

The silt that remained on the sandbars came not from the tributaries, but from lower down on the same bars. Eight days of high flows had been too many, flushing much of the sediment all the way out of the Grand Canyon. The Geological Survey concluded that the experiment had "not led to sustainable restoration and maintenance of sandbars in either Marble or Grand Canyon"; instead, "the canyons' sandbars continue to erode."[16] Overall, the controlled flood of 1996 had not only failed to restore sediment to the Grand Canyon, it had made sediment and habitat conditions worse. A great deal had been learned and in that sense the flood flows were a scientific success. But they did the creatures of the Grand Canyon no permanent good.

Scientists still believed the high flows could restore silt to the bars if they did not last as long and were more carefully timed to capture the maximum sediment input from the two tributaries. On 21 November 2004, 41,000 cfs jetted out of Glen Canyon Dam for two-and-a-half days. But afterward, the only sandbars that had gained sediment were those in the first thirty miles of the Grand Canyon.[17] The bars downstream had less sediment than before the experiment.

Almost a year after the November 2004 flood flow experiment, the Geological Survey published *The State of the Colorado River Ecosystem in the Grand Canyon.*[18] The massive report reviewed each aspect of Grand Canyon ecology, emphasizing the two attempts to rebuild the sandbars using flood flows. It made grim reading. Thirteen years after the Grand Canyon Protection Act had ordered their rescue, the endangered species in the Grand Canyon were worse off. Between 1990 and 2002, the number of humpback chub had fallen from 10,000 to 3,000. In a pitiful turn of events, the poor chub, having adapted over millions of years to survive where John Wesley Powell and his crew nearly starved, having had its numbers decimated by the operation of the dam and the invasion of nonnative rainbow trout, now suffered an infestation by an exotic species, the Asian tapeworm.[19] It would take a hard-hearted person not to wish well these beetle-browed, prehistoric survivors, who navigated the muddy waters of the ancestral Colorado River before our ancestors descended from trees. Humans have become the chub's worst enemy and its only hope.

On 12 February 2006, basing their case on the evidence presented in the Geological Survey report, five environmental organizations—Living Rivers, the Arizona Wildlife Federation, the Center for Biological Diversity, the Glen Canyon Institute, and the Sierra Club's Grand Canyon chapter—sued the Department of the Interior and the Bureau of Reclamation for violating the Endangered Species Act.[20] "We've been waiting patiently for 11 years," said the board chairman of the Center for Biological Diversity. "The Colorado River in the Grand Canyon is being destroyed, and the Glen Canyon Dam adaptive management program has been a failure." John Weisheit of Living Rivers accused the Bureau of Reclamation of remaining "hell-bent on destroying the Grand Canyon."[21] Instead of fighting the suit, the bureau agreed to settle. By 15 October 2008, it is to complete an environmental impact statement that will appraise the effect of the dam on the humpback chub and propose a long-term plan for protecting the species.

What might the bureau propose? The Geological Survey report offers three potential remedies: "Further restraints on daily powerplant operations, changes in monthly volume release patterns, or sediment augmentation."[22] The first two have already failed and in any case, when Lake Powell falls below the level of its power pool, the power plant will no longer operate and such restraints will no longer be possible. On the bright side, water leaving the dam through the outlet works then could be returned to its seasonal, pre-dam schedule: low in winter, rising steeply in late spring and falling in autumn. That would be a step toward restoring normalcy to the Grand Canyon, but the vital sediment would still be missing.

The bureau is studying the possibility of building pipelines to carry slurried sediment to the Grand Canyon from land deposits a few score miles away. The capital cost would be a few hundred million dollars and the annual operating costs in the range of $5 to $15 million.[23] These amounts are about the same as already spent in fruitless attempts to restore habitat in the Grand Canyon. The pipelines have a much better chance of working.

The Geological Survey report makes it clear that Glen Canyon Dam is driving species in the Grand Canyon to extinction. We can hope that the remedies that the Bureau of Reclamation will propose in 2008 will reverse the trend, but what if they do not? Then the United States Supreme Court will likely have to decide which has the higher priority—a dam or an ecosystem. As one of the river scientists put the dilemma, "Maybe we should tell the public that the cost of providing water and power is the decline of the Grand Canyon and the public will accept that. But the public might also say the Grand Canyon is too important and we should spend whatever it takes."[24]

In March 2008 the Bureau of Reclamation tried for the third time what had failed twice, releasing flood flows in an attempt to rebuild bars and beaches in the Grand Canyon. With Interior secretary Dirk Kempthorne on site, 41,500 cfs jetted spectacularly from the dam for sixty hours. In fall 2008, the bureau plans two more weeks of high flows. Until then, power demands in Las Vegas and Phoenix will once again dictate release schedules. And after the fall releases, the bureau plans no other high flows for the next five years. In the meantime, the chub will just have to fend for itself.

Surprisingly, a report released by the Geological Survey in December 2007 suggests that the best thing for the humpback chub may not be more water temporarily in the Grand Canyon but less in Lake Powell.[25] From 1989 to 2001, the number of chub in the Grand Canyon fell from nearly 10,000 to about 5,000. Curiously, since then the population has increased steadily to about 6,000 chub. What changed between 2001 and today? Lake Powell fell from full to less than half full. Could it be this simple: the less water in the reservoir, the more chub in the Grand Canyon? Though scientists are trying to understand the reason for the chub's partial recovery, they note that as the lake level falls, the warm surface is closer to the penstock intakes and the water exiting the dam and reaching the canyon below is warmer. In 2005, water temperatures in the mainsteam Colorado River near the Little Colorado were the highest since the reservoir filled in 1980.

By the time the Colorado River basin states and Mexican irrigators have taken their share today, they have sucked the Colorado River dry. South of the border, in the words of author Philip Fradkin, most of the time the Colorado is "a river no more."[26]

As a young man, pioneering ecologist Aldo Leopold canoed the vast and unknown delta of the Colorado River:

> I have never gone back to the Delta of the Colorado since my brother and I explored it, by canoe, in 1922. . . . For all we could tell, the Delta had lain forgotten since Hernando de Alarcón landed there in 1540. . . . On the map the Delta was bisected by the river, but in fact the river was nowhere and everywhere, for he could not decide which of a hundred green lagoons offered the most pleasant and least speedy path to the Gulf.[27]

The Leopold brothers paddled one of the great wonders of the natural world, a vast, fecund marshland covering almost two million acres. The source of the silt that sustained the delta lay hundreds of miles upstream, a mile high on the arid Colorado Plateau. Near sea level, ground to bits by its tumultuous journey and well-watered, the silt of the Colorado River Delta supported life of extraordinary diversity.

Alarcón had sailed up the Colorado in search of a Spanish obsession: the Seven Cities of Gold. Instead he met the Cucupá, who enjoyed a different kind of treasure. For them, the delta was close to paradise, a never-ending cornucopia of gifts from the gods brought south by the great, muddy river that never ran dry.

> The still waters were of a deep emerald hue. . . . A verdant wall of mesquite and willow separated the channel from the thorny desert beyond. At each bend we saw egrets standing in the pools ahead. Fleets of cormorants drove their black prows in quest of skittering mullets; avocets, willets, and yellow-legs dozed one-legged on the bars; mallards, widgeons, and teal sprang skyward in alarm. . . . At every shallow ford were tracks of burro deer. We always examined these deer trails, hoping to find signs of the despot of the Delta, the great jaguar, *el tigre*. We saw neither hide nor hair of him, but his personality pervaded the wilderness.[28]

As he floated the green lagoons, Aldo Leopold may have been aware that the Reclamation Service was developing plans that would eventually threaten the delta and endanger the Cucupá way of life. Sure enough, no sooner had Hoover Dam come on line than the flow of water across the border dropped sharply. Mexico could do little about it, for the Colorado River Compact had granted the nation rights only to the water left over after the Americans had taken as much as they wanted. The 1944 treaty was an improvement, guaranteeing Mexico 1.5 million acre-feet of Colorado River water at Morelos Dam, the lowest impoundment on the Colorado. Below Morelos, water enters the Canal Central and flows to the Mexicali and San Luis agricultural valleys, which together irrigate more acres than the Imperial Valley. Once the priority upstream became to fill Lake Powell, for two decades no water reached the delta of the Colorado River.

Then came the floods of the 1980s. Glen Canyon and Hoover dams had to "spill" water, anathema to reclamationists but life-restoring to the delta. From 1980 to 1993, a yearly average of nearly 5.5 MAF crossed the border; much of it reached and restored the delta. Again water filled the entangled oxbows, lagoons, and sloughs.[29] Native cottonwoods, willows, cattails, and reeds sprang to life. The recovery inspired Environmental Defense to publish *A Delta Once More,* describing the resurrection of the delta and how to sustain it.[30] The report showed that annual flows of only 32,000 acre-feet, supplemented once every four years by flood flows of 260,000 acre-feet—together less than 1 percent of the average flow of the river— would maintain a substantial portion of the delta. But as the twenty-first-century drought began, the flows to the delta once again dried up.

In 1993, in order to meet the terms of the 1972 agreement with Mexico, the Bureau of Reclamation completed a desalting plant at Yuma, Arizona. The agency designed the plant to process highly saline water from the Wellton-Mohawk Irrigation District and separate it into two streams. One would have a low enough salt content so that when added to the Colorado, it would lower the salinity of the mainstem. To prevent the brine stream from entering the river while it built the Yuma plant, the bureau dug a fifty-mile-long, concrete-lined canal from Arizona south to a point near the northeastern extremity of the Gulf of California. Storms damaged the intakes and prevented the plant from going into full operation, allowing the brine stream to continue to flow south down the canal. At 3,000 ppm, the water is too salty for crops but just right for vegetation in an estuarine marsh.

The salty water transformed a mud flat into a 50,000-acre wetlands called La Ciénega de Santa Clara, host to some four hundred wildlife

Figure 26. Desalting plant in Yuma, Arizona (Courtesy of Bureau of Reclamation)

species. As author Charles Bergman puts it, "The restoration was easy. No one had to do anything. Just add water."[31] On the other hand, the process could be as easily reversed: just remove water, or poison it. Were the Yuma plant to return to operation, only one-third as much water would reach the marsh and it would be three times as salty.[32] According to Ed Glenn of the University of Arizona, who has studied the marsh since 1991, if the Yuma plant were restarted, La Ciénega's "vegetation would disappear" in "selenium-laced, poisonous water."[33]

In June 2000, Defenders of Wildlife and other conservation organizations sued the Department of the Interior. The conservationists alleged that the bureau, the Fish and Wildlife Service, and the National Marine Fisheries Service had violated the Endangered Species Act by failing to consider the impact of the Lower Colorado River Multi-Species Conservation Program, referred to in chapter 14, on endangered wildlife below the border. In 2003, a federal court rejected the suit on the grounds that under the 1944 treaty, Interior has no authority to release additional water to Mexico. A few months later, an appeals court also rejected the conservationists' argument. Thus one avenue for getting water south to the delta—applying the Endangered Species Act across international borders—appears closed. Filing briefs against the conservationists and on behalf of the government agencies were the Arizona Power Authority, the Central Arizona Water Conservation District, the Coachella Valley Water District, the Colorado River Commission, the Imperial Irrigation District, the Metropolitan Water District of Southern California, the San Diego County Water Authority, the Salt River District, the Southern Nevada Water Authority, and the State of Arizona.

The Sonoran Institute believes that water from a source other than the Colorado River could still save the delta. If the price were right, farmers near the border might be willing to fallow their fields and sell their water rights. More water could come from captured irrigation runoff, whose most likely source is the Wellton-Mohawk Irrigation District. In its report, *Conservation Priorities in the Colorado River Delta,* the Sonoran Institute shows how little water it would take to save the delta and La Ciénega. But freeing up the water will require political leadership in both the United States and Mexico that so far has been seldom seen. United States water managers would have to agree to give up some of their precious share of the Colorado River in part to benefit environmental recovery in another nation. That would require a sea change in the way business is done on the river.

—

If there were any doubt, the desperate state of wildlife in the Grand Canyon and the deathbed struggle of the Colorado River Delta and La Ciénega de Santa Clara demonstrate that mega-dams destroy ecosystems. To prevent that outcome on the Colorado will require expensive and permanent remedies, yet to be discovered. If such remedies are not found soon, the Colorado River will become the site of a political and legal struggle that will make the battle over Echo Park look tame. The twenty-first-century contest will be but one of many fought as Lake Powell shrinks and the Law of the River breaks down.

PART FIVE

River of Tomorrow

EIGHTEEN

River of Law

THE COLORADO RIVER COMPACT of 1922 was the *sine qua non* for transforming the Great American Desert into today's vibrant Southwest.[1] By
guaranteeing each of the two basins 7.5 MAF, the compact gave the
Colorado River basin states the confidence to grow. Yet even if global
warming had never appeared, the compact contained the seeds of its own
eventual failure. Today we know that the river carries an average flow no
greater than 14.6 MAF. Yet, including the extra 1 MAF that the compact
allowed the lower basin, the agreement allocated a total of 16 MAF. Since
demand rises to consume supply, and since the compact overestimated
supply, it encouraged more development than the river can ultimately sustain, leading to an eventual crisis.

The compact also left fundamental questions unaddressed. How much
water was each basin state to receive? How much would go to Mexico?
How much might be owed Indian tribes? What do "consumptive use,"
"deficiency," and "surplus shall prove insufficient" mean exactly? Should
the flow of tributaries within a state count against its Colorado River
allocation? If the water was there and the upper basin needed it, could it
take more than 7.5 MAF? The most fundamental question—what to do
when the inevitable, long-term drought arrives—the compact left out.
The commissioners were far from ignorant of drought: they had lived
through more than one in the nineteenth century. They simply elected to

leave certain vexing issues—ones that they did not believe they had to solve or that were clearly insoluble at the time—for another group and another day. Had they attempted to address drought, Mexican rights, and Indian claims, there would have been no Colorado River Compact—or at least not then—and the future of the West would have unfolded differently. Two thorny issues would arise decades later—recreation and the environment.

Having grown up in nineteenth-century America, the commissioners would not recognize today's West. The dusty hamlets of their youth have transmogrified into giant, thirsty cities whose red roofs of Spanish tile spread ever outward like an amoeba, gobbling up land and water. That the driest states have become the fastest growing would have confirmed their belief in America's manifest destiny and in man's God-given obligation to make the desert blossom.

Over the half century that followed the compact, the Bureau of Reclamation built dams, reservoirs, canals, siphons, pumps, tunnels, and irrigation works of every sort on nearly every river in every western state. Because each project required its own enabling legislation, operating rules, and regulations, a largely unplanned legal superstructure arose. The resulting Law of the River is like a many-roomed mansion, each floor designed at a different time by a different architect. It still stands, first because the twentieth century was unusually wet and second because the upper basin developed slowly compared to the lower.

Just the *major* components of the Law of the River comprise a long list: after the compact itself, in chronological order they include the Boulder Canyon Project Act, the California Limitation Act, the California Seven-Party Agreement, the United States–Republic of Mexico Water Treaty, the Upper Colorado River Basin Compact, the Colorado River Storage Project Act, the *Arizona v. California* decision and decree, the Colorado River Basin Project Act, Operating Criteria for Colorado System Reservoirs, Minute 242 to the U.S.–Mexico Treaty, the Colorado River Basin Salinity Control Act, the Grand Canyon Protection Act, the Interim Surplus Guidelines of 2001, and the Colorado River Water Delivery Agreement of 2003.[2] Also applying, though to what extend the courts will have to tell us, is the environmental legislation of the 1960s and 1970s: the National Environmental Protection Act, the Endangered Species Act, and the Clean Air and Clean Water Acts. Experts—and some attorneys and scholars devote their careers to the Law of the River—agree that it comprises at least fifty different governing documents.

Suppose that inertia, self-interest, and fear of coming out on the losing end of a new agreement prevent the basin states from revising or replacing the Colorado River Compact. Science-based modeling assumptions show that rising demand and shrinking supply will deplete the Colorado River system until there is water enough for only one large reservoir—and some of the time that reservoir will be no more than half full. As the hydrologic system falters, how might the legal system respond? Since we do not know how much water the river will carry from year to year, or how to predict human behavior, there is no way to answer the question precisely. What we can do is explore one possible future out of many, letting our imagination roam, and let that one stand as a warning. For this journey into the future, we can use the CROSS spreadsheet model described in chapter 14.[3]

The model requires us to choose a start year from the historic flow record. If we select 1982, the flood years of 1983 and 1984 would quickly follow and the time of possible system failure would be displaced farther ahead. If instead we choose the year the Dust Bowl began, the results would soon reflect the 1930s drought, with the opposite result. Since the purpose of this exercise is in part to sound a tocsin, let us pick a year that will give results closer to the worst case than to the best. The midpoint of the twentieth century, 1950, will serve well. The drought of 1953–57 soon followed, but several other start years in the historic record give worse results.

Let us begin our imaginary future at the end of October 2007. Lake Powell and Lake Mead each held 49 percent of capacity. A key assumption of the model is that by 2050, global warming will reduce the flow of the Colorado River by 10 percent.[4] That amounts to 1.4 MAF, a significant loss on a river with no surplus. Even so, the basin states could accommodate the shortfall by increasing conservation, limiting development, transferring water from agriculture to cities, and the like. But will they do so voluntarily? Not all of them may, for some states have less of an incentive to adapt than others. Reflecting the provenance of western water law in the doctrine of prior appropriation and in the rural character of the 1920s West, the Law of the River favors some over others: the lower basin over the upper, agriculture over cities, California over Nevada—and especially, California over Arizona. Human nature being what it is, the states with the least to gain from abandoning business as usual will be the last to do so.

The model forecasts that between 2008 and 2010, corresponding to the relatively wetter years from 1950 to 1952, Lake Powell gains volume. Across the West, the water establishment notes the flowing creeks and rivers, the rise of the reservoirs and the shrinking of their bathtub rings, and breathes a collective sigh of relief. Conservative pundits note with glee that climate change is turning out not to be nearly as bad as scientists had predicted.

The wet years encourage the State of Utah to press ahead with its plan to build a $500 million, 158-mile pipeline to bring 70,000 acre-feet of water from Lake Powell to burgeoning St. George in the Beehive State's southwestern corner.[5] At 335 gallons per person per day, the red-rock community has one of the highest per capita water consumption rates in the country. The water from Lake Powell will allow St. George to grow from the 120,000 it held in 2005 to 480,000 by 2050.[6]

Up where Utah, Colorado, and Wyoming meet, entrepreneurs had planned to build a $4 billion pipeline to take water from Flaming Gorge Reservoir on the Green River above Dinosaur National Monument to Denver and other Front Range cities.[7] The pipeline would carry up to 450,000 acre-feet per year at a capital cost of nearly $9,000 per acre-foot. But California sues and halts the transfer.

As the level of Lake Powell fell in the first few years of the twenty-first century, the Navajo Nation had begun to worry. Pumps send cooling water from Lake Powell to the Navajo Generating Station near Page, Arizona, built in place of power dams in the Grand Canyon. But the intakes are at the same elevation as the reservoir's power pool. If the lake were to fall below that level, which, as the dry years return, is beginning to look likely, no water could reach the generators and the station would have to shut down. That would close off an important source of jobs and revenue to the economically depressed reservation. The Navajo urge the consortium that manages the plant to build new intakes at Lake Powell's dead pool elevation—the reservoir cannot fall lower than that.

As a second strategy, the Navajo Nation lodges a claim in federal district court in Arizona to present "perfected" rights of 1,000,000 acre-feet of Colorado River water annually. The legal term refers to water rights that predate, and thus are senior to, any later claim. Since the decision in *Winters vs. United States* in 1908, a succession of court decisions have confirmed that a tribe's right to water dates to the time the government established its reservation. In their suit, the Navajo point out that the precedent establishes their priority to 1868, forty-four years before Arizona became a state. Based on the rules that various courts have established, the

Navajo demonstrate that they could claim as much as five million acre-feet of Colorado River water. But, recognizing the reality of global warming, the tribe magnanimously limits its claim to one-fifth the amount it could demand. The suit presents the secretary of the Interior, overseer of the Bureau of Reclamation, with a potential conflict of interest. Federal law entrusts the government with responsibility for defending the rights of Indians and protecting their tribal lands and resources. If the government fails to do so, a tribe has an important legal platform—breach of trust— from which to sue. The Navajo Nation has used this platform before and will certainly do so again where water is at stake.

Las Vegas takes its share of Colorado River water from two intakes into Lake Mead, one at the 1,050-foot elevation and the other at 1,000 feet. The bureau has given assurances that it will not allow the reservoir to fall below the lower of the two intakes. But the general manager of the Southern Nevada Water Authority well remembers that in early 2005, Lake Mead had only another 80 feet to fall before it dropped below the higher of the two intakes. The authority speeds up its plan to build a new pipeline into the reservoir at 950 feet, 55 feet above Lake Mead's dead pool. This "third straw" will cost $650 million, part of a $7.6 billion set of construction projects that will allow Las Vegas to more than double its water consumption.[8] Should Lake Mead ever fall below 950 feet, it could not supply Las Vegas with water. But in its final environmental impact statement back in 2007, the bureau had forecast the probability of Lake Mead falling below about 1,020 feet as less than 10 percent, so there is no need for worry.[9]

Nevada had earlier threatened to capture 100,000 acre-feet from the Virgin River, which flows south from its headwaters in Zion National Park and into Lake Mead.[10] When the proposal first appeared, Colorado objected on the grounds that tributary water, even though drawn from Lake Mead, ought to count against Nevada's entitlement to 300,000 acre-feet of Colorado River water.[11] If Nevada takes 100,000 acre-feet from the Virgin, Colorado now argues, Nevada should receive 100,000 less from the mainstem. After all, even though the Colorado River Compact is vague about how to count tributary flows, upper basin tributaries like the Gunnison River in Colorado are considered part of the mainstem. Why should lower basin tributaries be treated differently? Colorado and the other three upper basin states sue in federal district court to prevent Nevada from proceeding with its plan to draw on the Virgin.

Next, the Southern Nevada Water Authority comes up with a scheme to benefit from desalting, a seemingly impossible dream for Las Vegas, located hundreds of miles from seawater. But never underestimate the invention that a great thirst can spawn. The authority will pay for a desalting plant on the Sea of Cortés that will supply fresh water to nearby border cities like Mexicali. In return, Mexico will cede to Nevada the same volume of water from its share of the Colorado River. Following Nevada's lead, Phoenix and Tucson plan their own Mexican desalting plants, intending to swap the water. These plans assume that desalting will produce a significant, dependable amount of fresh water and that in a prolonged drought, nations and water districts will be willing to swap some of their Colorado River allocation for it.

The intakes to Glen Canyon's turbines are at 3,470 feet, but when the reservoir falls to 3,520 feet in late 2012, whirlpools and eddies form on Lake Powell's surface, sucking air into the precisely machined turbines and causing them to shudder and vibrate. If the shaking continues, the turbines could break loose and destroy the powerhouse. Managers have no choice but to close the generator intakes, ending power production at Glen Canyon.

The highest priority of the Law of the River is that Glen Canyon Dam release 8.23 MAF annually. To supply its obligation in 2010, the upper basin has to draw down its own reservoirs. The withdrawals amount to only 40,000 acre-feet, but by 2012 they rise to 50,000 acre-feet and by 2013 they climb to over 100,000 acre-feet. Rather than create a crisis, the upper basin accedes to the reductions but puts its lawyers to work.

With the Law of the River unrevised, Glen Canyon Dam continues to release more water each year than Lake Powell receives. By 2022, the releases plus evaporation losses have dropped Lake Powell to 2 MAF, barely above dead pool. Hydrologists forecast that in order to deliver its obligation the next year, the upper basin will have to cut back its consumption by nearly 3 MAF, more than half its planned draw on the Colorado River. Rather than suffer the regional recession that such cutbacks would cause, the basin seeks an injunction to allow it to maintain consumption. The upriver states point out that the Colorado River Compact does not require them to *deliver* water to the lower basin. Rather their duty is not to "cause the flow of the river at Lee Ferry to be depleted." Natural drought and global warming, acts of God and man respectively, rather than consumption, have depleted the river, so the upper basin claims. With Arizona, California, and Nevada plus each state's irrigation districts filing an opposing brief, the district court declines to grant the injunction. The upper basin appeals, but experts on the Law of the River agree that the Supreme

Court will have to decide. If *Arizona vs. California* is a guide, that will take years if not a decade or more. Meanwhile, with no way to make up the missing water, the economy of the upper basin goes into a tailspin.

By the early 2020s, Lake Powell is effectively at dead pool and the two reservoirs together hold less than 10 MAF. Since Glen Canyon Dam no longer generates power and since low water has compromised Lake Powell's recreational appeal, the basin states now decide to amend the Law of the River. From here on out, Glen Canyon Dam will release all the water it can and none will be held back in an effort to salvage Lake Powell. Except for a few wet periods, for the rest of the century Lake Powell remains at dead pool.

For some constituencies, the fall of Lake Powell has its positive side. Water exiting through the dam's outlet works at Glen Canyon now comes from near the reservoir's warm surface. The Colorado River in the Grand Canyon again ebbs and flows on its natural schedule. Slurry pipelines from Navajo Canyon carry sediment to the Grand Canyon each year. The canyon's ecosystem now has the three ingredients it needs for good health: warm, silty water that rises and falls on Nature's timetable. Upstream, for the first time in sixty years, rafters are able to float the top forty miles of Glen Canyon. The little town of Page, Arizona, converts its economy from boating and fishing to rafting and ecotourism.

As the upper basin draws down its reservoirs, western water law requires it to work back through the list of project priority dates, shutting them off in reverse chronological order. Since city water projects generally are last in, they tend to be first out. Take Albuquerque. For decades the city relied on an aquifer for its water supply, withdrawing twice the volume that Nature replenished. As a result, the water table fell by 180 feet, compelling the city to look for other sources. In the 1960s and 1970s, Albuquerque and the Bureau of Reclamation came up with a plan to divert 50,000 acre-feet from the San Juan River to the north and send it through tunnels and a series of reservoirs down to the city, where a $160 million plant would purify the water and supply 70 percent of Albuquerque's needs. Instead, as one of the last projects to come on line, the San Juan diversions are among the first to close, forcing Albuquerque to fall back on its dwindling groundwater supplies.

St. George completes its pipeline from Lake Powell in 2015, but as the last project dependent on upper basin water, it is one of the first to be shut down. The city stops issuing new water hook-ups, leaving three new subdivisions high and dry. Farther upstream, the Colorado–Big Thompson project, which postdates the Colorado River Compact and therefore has a lower priority, also shuts down. Diversions from Colorado–Big Thompson to Front Range cities drop from nearly 500,000 acre-feet to 70,000, drastically cutting water supplies to Denver, Boulder, Colorado Springs, and the other cities east of the Rockies.[12] The upper basin sinks into a depression not seen since the 1930s.

Throughout these years, the Colorado River Compact has been more of a hindrance than a help—only one section even mentions the possibility of drought. Article IIIc states that when too little surplus water is available to meet the 1.5 MAF obligation to Mexico, "the burden of such deficiency shall be equally borne by the upper basin and the lower basin, and whenever necessary the states of the upper division shall deliver at Lee Ferry water to supply one-half of the deficiency."[13] By the 2020s, not only must the upper basin cut consumption and damage its economy to benefit the lower basin states, but if it is to meet the United States treaty obligation, it would have to cut consumption even further and deepen its economic depression. The basin balks, arguing that it is unable to make up its half of the debt to Mexico because it has no surplus water and in any case, did not cause the deficiency. In its Supreme Court brief, the basin argues that the compact and all subsequent treaties and legislation are null and void because first, the Colorado River Compact overestimated the amount of water in the river and second, for years the Bureau of Reclamation failed to take scientific evidence into account, misleading the public and state water officials into believing that water shortages in the Colorado River basin were next to impossible.

The 1944 treaty with Mexico covered not only the Colorado River; it required Mexico to deliver to the United States 350,000 acre-feet of water each year from the Rio Grande and its tributaries.[14] Over the 1990s and early 2000s, Mexico fell behind and by 2002 its debt on the Rio Grande reached 1.5 MAF. A wet 2005 allowed Mexico to work off the backlog, but as the years pass, the debt rises again and by 2020 it exceeds one million acre-feet. Mexico announces that it is unable to provide water from the Rio Grande due to "extraordinary drought," a term used but not defined in the 1944 treaty, and that it will not make up its debt on the Rio Grande until the United States does likewise on the Colorado. When the United States responds that it does not have the water, Mexico sues in the World Court. Meanwhile, reservoir evaporation and lower flows have raised the salinity of water in the lower Colorado River above the 1972 treaty limit. Mexico files a second suit, adding water quality to its complaint over quantity.

In 1968, to gain passage of the Central Arizona Project, Arizona had to agree to subordinate its water rights to those of California. Any shortfall in the lower basin Arizona will have to make up from its own share of the Colorado. Fighting a two-front war against Indians and Californians, Arizona grows desperate. Back in 2007, at the urging of the secretary of the Interior, the seven basin states had come up with a joint plan to address water shortages on the Colorado. In an extreme drought, the lower basin agreed to take a reduction of 600,000 acre-feet, almost all of it to come from Arizona's share. As long as such a drought turned out to be temporary, the withdrawal would have little effect on Arizona's actual use of water, for the state had come up with a clever scheme of its own. In 1999, Arizona gained government approval to take more Colorado River water than it needed and store it in underground aquifers. By 2006, Arizona had banked over 2 MAF. But by the mid-2020s, years of drought have emptied the bank—yet the Law of the River still obligates Arizona to make up any deficit in the lower basin.

Arizona seeks relief in district court, arguing that it had agreed to make its rights junior to California's only because the Central Arizona Project legislation promised that one day the Colorado River would be augmented annually by 2.5 MAF. From where the augmenting water would come, the project's language did not specify. Various schemes had been proposed to conjure up the water, but none has brought a drop to the Grand Canyon State. Cloud seeding has been a bust and desalting has produced only a relative trickle. The only way to augment the Colorado River is to transfer water from outside its basin, but with the entire West facing the same permanent drought, no state will agree.

The court rejects Arizona's claim, leading the eight-member Arizona delegation in the House of Representatives to introduce legislation to repeal the sections of the 1968 Colorado River Basin Project Act (which includes the Central Arizona Project) that subordinate Arizona's water rights to those of California. The fifty-three-person California delegation responds with its own bill to rescind both the Colorado River Compact and the 1948 Upper Colorado River Basin Compact and return the West to the doctrine of prior appropriation.[15] With a former California governor in the White House and her party in control of Congress, the bill stands a good chance of success.

Like other cities and irrigation districts, Phoenix assumed that droughts would develop slowly and eventually end—after all, they always had.

The city's 2005 water plan noted that climate change could trigger short-ages but, finding it "difficult to predict the exact impact," did not include the effects of global warming in its modeling.[16] In the first decade of the twenty-first century, western water agencies from the Bureau of Reclamation on down excused inaction on the grounds that scientists had been unable to agree on precisely how much global temperatures would rise and how much runoff would fall.

Though Phoenix receives only seven inches of rain per year, by relying on groundwater, supplies from the Salt River Project, and in more recent years, from the Central Arizona Project, its metropolitan area grew from 330,000 in 1950 to four million in 2007. By 2020, the population has risen past six million and the central Arizona megalopolis has a larger area than New Jersey. Though conservation measures have lowered per capita water consumption, each central Arizona resident still consumes an average of 210 gallons per day. As releases from Hoover Dam have declined, as the Salt River reservoirs have dried up, as the water bank has emptied, Phoenix has no choice but to return to groundwater.

During the boom years, central Arizona drew down its groundwater supplies so rapidly that in some areas the water level fell 300 to 400 feet.[17] In 1980, an alarmed state legislature passed a groundwater reform bill that required developers to show in advance of construction that they had an assured supply of surface water that would last one hundred years. The obvious source of the water was the Central Arizona Project, but develop-ers often wanted to build where it was impractical or overly expensive to bring project water. Then, as always, water began to flow through a loop-hole and uphill to money. Developers were allowed to build and continue to draw on groundwater, as long as the eventual homebuyer paid to have the same amount of surface water taken from somewhere and restored to the underground reservoir.[18] By 2020, no somewheres are left.

Western temperatures rise even higher than the top of the range forecast by climate scientists. By 2020, the immense mass of asphalt and concrete that make up the Phoenix heat island hold average summer nighttime temperatures at 95 degrees; on some nights the temperature never drops below triple digits. June and July daytime temperatures average 119 degrees. On 1 August 2020, Sky Harbor Airport registers a daytime high of 133 degrees and a nighttime low of 102 degrees. With demand for power far exceeding supply, lights dim and air conditioners shudder and fail. High-ways and airport runways buckle; rails expand and warp as they did during the Dust Bowl years. The summer heat kills 1,000 people, most of them homeless or aged. While the Arizona sun is above the horizon, life in Phoenix is a mad dash from one air-conditioned refuge to another. During daytime hours, children and pets stay indoors.

The acreage and duration of wildfires quadrupled from the mid-1980s to 2007; by 2020, both have doubled again. Across the American West of the 2020s, smoke from wildfires is a constant companion. And in Phoenix, smoke is not the only substance in the air. Also present is a fine dust, some blown all the way from China, some coming from the Salton Sea west of Phoenix. Overflow of the Colorado River in the 1905–06 floods created the Salton Sea; agricultural runoff had since kept it alive. But as the amount of water used for irrigation declined, the sea has dried up. Its mud bottom holds enough arsenic, selenium, heavy metals, and pesticides to qualify as a hazardous waste site. Now, with the seabed a bone-dry desert, the toxic particles blow straight for Phoenix, only two hundred miles downwind. The smoke from forest fires and the dust from the Salton Sea mix with the brown smog that had begun to afflict Phoenix in the 1990s to create a daily air-quality emergency. The city whose dry, pristine air drew asthmatics and others with respiratory problems now drives them away—if they can afford to leave. But as in New Orleans during Hurricane Katrina, many of the aged and infirm have nowhere to go and no way to get there.

The World Court upholds Mexico's claim and requires the secretary of the Interior to drop the level of Lake Mead below its power pool, ending electricity generation at Hoover Dam. With both Glen Canyon and Hoover dams offline, electricity prices rise across the West, and brownouts and power outages become the rule. Nuclear plants around the country were supposed to stave off the energy crisis and power new desalting plants, but delay in getting the nuclear plants approved and built leaves them years away from production. The problem of what to do with nuclear waste remains unsolved and the controversy over using Yucca Mountain, Nevada, for that purpose now enters its forty-second year.

With both surface and groundwater supplies severely limited and no relief in sight, Phoenix declares a stage-four water emergency, its highest level. The state legislature rescinds the Groundwater Management Act. Voluntary reductions having long since failed to conserve enough water, Phoenix enforces rationing. Watering lawns, washing cars, and splashing in water parks are distant memories. The two hundred golf courses in Phoenix and Scottsdale have been closed for years, their verdant fairways and manicured greens blown away on the hot, dry wind. Valves attached to water meters automatically shut off the flow when consumption exceeds the limit. Armed water police with the authority to shut off valves and make arrests patrol neighborhoods. Phoenix doubles the price of water to residences, raises it even more for the heaviest users, and prohibits new water hook-ups. Home construction shuts down and the once-booming central Arizona real estate market collapses. As tax revenues decline, Phoenix runs short of funds and rating agencies reclassify its bonds as junk.

Following Nevada's example, Phoenix begins to build a desalting plant on the Sea of Cortés. But as the border crisis intensifies, and with its own water supplies at dangerous lows, Mexico nationalizes all American-owned factories in the country, including the desalting plants and the maquiladoras. By the 2020s, with water, the stuff of life at stake, it is every nation for itself.

Businesses and families begin to abandon Phoenix, creating a *Grapes of Wrath*–like exodus in reverse. Long lines of vehicles clog the freeways, heading east toward the Mississippi and north toward Oregon and Washington. Burning hot, parched, and broke, the city that rose from the ashes achieves its apogee and falls back toward the fire.

NINETEEN

The West against Itself

FOR THE COLORADO RIVER basin and the Southwest, the threat from global warming lies not in the comfortably distant future—the threat is here today. West of the 100th meridian, the danger derives not from the slow rise of the sea but from the much more rapid fall of the reservoirs. By itself the West can do nothing to reduce the extent of warming on a global scale. No matter how much the region might cut its own greenhouse gas emissions, the reduction will have no worldwide effect: western temperatures will continue to rise; snowpack and runoff will continue to decline. What the West can do is concentrate on mitigating the effects of climate change. It can start now, without waiting for more data and precise forecasts.

First, of course, the region and its leaders would have to accept that business as usual cannot continue. An article that appeared just as this book was being completed suggests that such acceptance may be farther in the future than even a pessimist would have imagined. "Developers plan massive water park in Arizona desert; project would use 100M gallons a year," read the headline. Mesa, Arizona, an eastern suburb of Phoenix, has no property tax and as a result its growth has lagged behind that of other central Arizona cities. To bring in needed revenue, on 6 November 2007, Mesa voters overwhelmingly approved $35 million in tax breaks to developers who would build a water park to provide the desert community with surfing, snorkeling, scuba-diving, and kayaking. Unlike the fountains in

Las Vegas, the Mesa water park could not use recycled water. Instead it would draw upon groundwater that contains unsafe levels of arsenic and build a plant to remove the poisonous chemical. One of the project developers inadvertently captured the way in which the West can be its own worst enemy: "I couldn't imagine raising my kids in an environment where they wouldn't have the opportunity to grow up being passionate about the same sports I grew up being passionate about."[1]

Once the Colorado River basin states recognize that what worked in the last century will not work in this one, their first step will be to replace the Colorado River Compact. The agreement is a construct of men who knew little of the river's long-term flow and who had no way to foresee the population explosion of Phoenix and Las Vegas, for example, much less global warming. In spite of its deficiencies, the compact has served well for eighty-five years, allowing a remarkable civilization to rise in a desert. But with demand for Colorado River water already consuming the supply, and with supply having already started to fall, the compact will soon become obsolete, and then dangerous. To prevent some version of the legal paralysis imagined in chapter 18, the basin states and Mexico urgently need to forge a new agreement. Given how much more is at stake today than in the 1920s, reaching agreement on a new compact will be much more difficult. A new Colorado River commission should comprise all the major stakeholders, including representatives of Native American tribes, Mexico, environmental organizations, and others. To break the habits of a century and rise above vested interests will require more leadership on the Colorado River than has been seen so far, but without it, the people of the basin will be bound to suffer. And they do not have to, for if only the West used water differently, it has enough to buy time while it adjusts to the new climatology.

Agriculture now uses 80 percent of western water. Of the remaining 20 percent, cities may use half for landscaping. And what do western landscapers plant? Until recently at least, the same water-gobbling grasses, trees, and shrubs that their eastern ancestors planted. Suppose the people of the Colorado River basin were to take two steps. First, reduce the percentage of water going to agriculture from 80 to 70 percent. Second, reduce the amount of city water going for landscaping from, say, 50 percent to 5 percent. How much water would these two steps free up? About 2.7 MAF each year, nearly 20 percent of the annual flow of the Colorado River.

But would not a reduction in the amount of water for agriculture in the West deprive Americans of the fresh produce they need—winter vegetables, for example? No, because much western acreage goes to feed cattle, not people. In California, source of a large fraction of the nation's year-round produce, 20 percent of agricultural water grows alfalfa. Western agribusiness land that does not grow food for cattle is apt to be planted in heavily subsidized cotton and rice, thirsty crops better grown somewhere other than a desert.

One way to encourage the transfer of water from agriculture to cities is to end government handouts to agribusinesses and require them to pay their way. In the Westlands Water District in California's western San Joaquin Valley, for example, one family farm of 29,000 acres receives an annual taxpayer-provided water subsidy estimated at $3.5 million.[2] Other government agencies provide even more subsidies. According to the Environmental Working Group, between 1995 and 2005 the United States Department of Agriculture gave California agribusinesses subsidies worth nearly $6 billion.

Government can facilitate markets that allow water basins, states, irrigation districts, cities, and others to trade water. In the Murray-Darling river basin, Australia has not only ended subsidies, it has encouraged trading and found that water moves to higher value, less water-intensive crops.[3] The West has begun to experiment with water transfers, as in the swap between the Imperial Irrigation District and San Diego. Much more will have to be done.

Beyond changing how it allocates water, conservation and greater efficiency can save the West millions of additional acre-feet. California's Water Plan 2005 estimates that greater use of recycled water would yield the state 1.4 MAF annually. The plan also found that more aggressive urban conservation could increase supply by over 3 MAF per year.[4] California has already shown it can adopt such measures: the state used less total water in 2001 than in 1975, even though population rose by 60 percent and the gross state product more than doubled.[5] The same methods can free up millions of acre-feet in every western state and provide breathing space while the West adjusts to a new climate regime. But if the new supply merely drives more development, as it always has done, the West will lose a one-time opportunity that an unforgiving Nature will not provide again.

Revising the Colorado River Compact and changing the way the West uses water will not be the first choice of the western water establishment. Instead, we already hear that the solution to less water is more dams. Before accepting that equation, Americans ought to consider the lessons of the last century. One is that few acceptable dam sites are left in the West. At least one dam, and sometimes a dozen, block nearly every western river. The sites that remain are the least safe and would store the least water at the highest cost—otherwise they would already have dams. Second, experience in the Grand Canyon and around the world teaches that dams destroy ecosystems. Dams and the benefits they provide are temporary; extinction is forever. When the last hump-backed chub is gone, there will never be another in this or any universe. If society wants more dams, it will have to accept fewer species. Third, new reservoirs will eventually silt up; by building them one generation foists off the need to find sustainable solutions onto its children and grandchildren.

Two new reservoirs proposed for California make the case against new dams. The Sites Reservoir northeast of San Francisco would not even impound a river. Instead, pumps would lift water from the Sacramento River and send it to the new reservoir via canals and irrigation works. The Temperance Flat Dam on the upper San Joaquin River would drown existing power plants and destroy more hydropower than it generates. Temperance Flat would inundate 7,000 acres of public land in the San Joaquin Gorge, threatening scores of sensitive species. The reservoir might not even fill: the San Joaquin is already so dammed and diverted that in most years it runs dry downstream of Fresno.

From listening to dam boosters, one gets the impression that the 500,000 acre-feet of water that the two new California reservoirs would release each year—if they fill—would be new water to relieve future shortages. But except for the potential contribution of desalting plants, yet to be realized, there neither is nor ever will be any new water. There never has been. On the earth's surface today, appreciably no more fresh water exists than did a million years ago. Dams do not create water; all they can do is gather and manipulate water already present. And each year evaporation costs reservoirs several percent of their storage. In the face of a shortage that may last for a century or more, dams waste water and money and delay action that might actually do some good.

What then are we to make of the claim that new dams in California and elsewhere will "combat climate change"? There is only one way in which they might and that is far from certain. As we have learned, global warming is already causing alpine snowfields to melt sooner. If existing reservoirs cannot contain the earlier spring snowmelt surges, water will spill over the dams and cause downstream flooding. But western reservoirs today have plenty of room. Lakes Powell and Mead are less than 50 percent full and California's reservoirs are about 80 percent full. Since global warming will reduce runoff and deplete them further, there is a good chance that western reservoirs will be able to contain the spring snowmelt surges. On the first of April 2007, the Sierra snowpack was 40 percent of average, the lowest it had been in nearly two decades.

The claim that new dams will combat climate change is the latest in a long list of fictions. Start with the International Irrigation Congress in 1893 and the irrigation boosters who booed John Wesley Powell. Major Powell's scientific calculations were wrong, they argued: western rivers and streams hold enough water to irrigate all western lands. Congress passed the Newlands Act of 1902 explicitly to benefit the small farmer who would live on and till his own land, not to exceed 160 acres. Newlands himself described what would happen otherwise: "Should the government grant these lands to corporations, it would absolutely abrogate the trust it holds for the entire people. No greater outrage could be effected."[6] But in one of its first projects, on Arizona's Salt River, the Reclamation Service began to betray its founding principles. Within two decades, it had betrayed them all. Meeting in 1922, the Colorado River commissioners allocated at least 1.5 MAF more water than the Colorado River carries. A year later, the Coolidge administration's fact-finding commission complained that the cost estimates of the Reclamation Service were "much lower than was justified," a finding amply confirmed in each subsequent review. Congress passed the Colorado River Storage Project in 1956 even though Wayne Aspinall, the Bureau of Reclamation, and many others knew that the river carries substantially less water than the commissioners allocated; even though Luna Leopold showed that a dam at Glen Canyon was unnecessary to regulate the river; and even though Senator Paul Douglas and others demonstrated that the CRSP was a poor use of taxpayer dollars—a boondoggle, he called it. In the 1968 Colorado River Basin Project Act, Arizona agreed to subordinate its rights to California's on the promise that 2.5 MAF of water from some unspecified and to date unidentified source would augment the Colorado River. Jump ahead to the twenty-first century and find the Bureau of Reclamation using the wettest of the last five centuries as the foundation for its planning, while avoiding climate change science.

During a century of reclamation, the West accepted these and other fictions. It built Glen Canyon Dam, a magnificent engineering achievement, and created Lake Powell, as beautiful in its way as any natural lake. The pair stand as emblems of the century of reclamation and of man's ability to control Nature—for a time. Now a new climatology, one that derives from man's own actions, threatens that control. We can save either Lake Powell or Lake Mead, but not both. Can we save the desert civilization that they symbolize? As Bernard DeVoto put it sixty years ago, "The future of the West hinges on whether it can defend itself against itself."[7]

ACKNOWLEDGMENTS

FRIENDS AND COLLEAGUES WHO REVIEWED all or parts of various drafts include Lloyd Carter, Niklas Christensen, John Dohrenwend, Karl Flessa, Owen Lammers, David Morrison, Donald Reich, Ernest Smerdon, Judi Stauffer, Dave Wegner, John Weisheit, Flake Wells, Connie Woodhouse, and Donald Worster. The book would not have been possible without the fine assistance of librarians at the University of California at Santa Barbara and at the University of Southern California. Alan Barnett and the Utah State Historical Society kindly copied pages from a hard-to-find report. Agent John Thornton and editor Blake Edgar gave encouragement and expert advice. Matthew Winfield of the University of California Press assisted admirably; Kathleen MacDougall did a world-class job of editing.

This is a good place to acknowledge that the phrase "river of empire" is borrowed from Don Worster and the phrase "river of controversy" from Norris Hundley, Jr., both masters of western history.

My wife, Joan, not only read the manuscript carefully but patiently put up with an obsessive writer. To her goes the dedication.

NOTES

CHAPTER ONE

1. Stephen E. Latham, *Glen Canyon Dam, Arizona: Dam Failure Inundation Study* (Denver: Bureau of Reclamation, 1998), pp. 7–9.

2. Richard West and Richard Meyer, "Glen Canyon Dam Groans under Flooding," *Los Angeles Times,* 1 July 1983.

3. T. J. Wolf, "How Lake Powell Almost Broke Free of Glen Canyon Dam," in *Water in the West: A High Country News Reader,* ed. Char Miller (Corvallis: Oregon State University Press, 2000), p. 87.

4. Ibid., p. 89

5. L. B. Leopold, "Oral Interview" (Bancroft Library, University of California at Berkeley, 1993), p. 181.

6. Michael C. Robinson, *Water for the West: The Bureau of Reclamation, 1902–1977* (Chicago: Public Works Historical Society, 1979), cover photograph.

7. Richard Meyer, "Raging Colorado Flood Batters Grand Canyon," *Los Angeles Times,* 25 June 1983.

8. James Udall, "When You See a River as Plumbing, How Can You Turn It Off?" *Los Angeles Times,* 1 July 1983 (italics in original).

9. Michael Patrick Ghiglieri, *Canyon* (Tucson: University of Arizona Press, 1992), p. 6; Udall, "When You See a River as Plumbing, How Can You Turn It Off?"

10. Meyer, "Raging Colorado Flood Batters Grand Canyon."

11. Steven Gloss, Jeffrey E. Lovich, and T. S. Melis, "The State of the Colorado River Ecosystem in the Grand Canyon," U.S. Geological Survey Circular 1282 (2005).

12. John S. Weisheit, and S. M. Fields, "The Moab Mill Project" (Living Rivers, 2006).

13. W. L. Rusho, "Bumpy Road for Glen Canyon Dam" (paper presented at the symposium on the history of the Bureau of Reclamation, University of Nevada at Las Vegas, 2002), p. 17 (Rusho's italics).

14. David Wegner, personal communication.

15. Wolf, "How Lake Powell Almost Broke Free of Glen Canyon Dam," p. 90.

16. Robert A. Jones, "Worst of Flooding Over, Watt Declares," *Los Angeles Times,* 2 July 1983.

17. Wolf, "How Lake Powell Almost Broke Free of Glen Canyon Dam," p. 91.

18. Steven Hannon, "The 1983 Flood at Glen Canyon" (2003), http://www.glencanyon.org.

19. Tom Knudson and Nancy Vogel, "Aging Dams, Already under Siege, Face New Pressures," *Sacramento Bee,* 25 November 1997.

20. Bureau of Reclamation, "Challenge at Glen Canyon," Open Video Project (1996).

21. Udall, "When You See a River as Plumbing, How Can You Turn It Off?"

22. Ibid.

CHAPTER TWO

1. Robert A. Young, ed., "Coping with Severe Sustained Drought on the Colorado River," *Water Resources Bulletin,* vol. 31 (American Water Resources Association, 1995).

2. Wallace Stegner, "Glen Canyon Submersis," in *The Glen Canyon Reader,* ed. Mathew Barrett Gross (Tucson: University of Arizona Press, 2003), p. 138.

3. Bradley Udall, "Climate Change Predictions and Impacts on Colorado River Water Supplies" (paper presented at the Upper Colorado River Commission, Las Vegas, 13 December 2006).

4. R. Seager, M. Ting, I. Held, Y. Kushnir, J. Lu, G. Vecchi, H. Huang, N. Harnik, A. Leetmaa, N. Lau, C. Li, J. Velez, and N. Naik, "Model Projections of an Imminent Transition to a More Arid Climate in Southwestern North America," *Science* 316 (2007): 1181–84.

5. Professor Dan McCool and others have used the term "Iron Triangles" to refer to certain actors in the drama of twentieth-century reclamation. I use "Concrete Pyramid" to refer to the quadrumvirate that dammed the American West: the Bureau of Reclamation, irrigators, construction companies, and politicians.

CHAPTER THREE

1. T. Jacobsen and R. M. Adams, "Salt and Silt in Ancient Mesopotamian Agriculture," *Science* 128 (1958): 1252–57.

2. Ibid., p. 1257.

3. Ibid., p. 1252.

4. Ibid., p. 1257.

5. Sandra Postel, *Pillar of Sand: Can the Irrigation Miracle Last?* (New York: W.W. Norton, 1999), pp. 25–26.

6. Ibid., pp. 26, 27.

7. Donald Worster, *The Wealth of Nature: Environmental History and the Ecological Imagination* (New York: Oxford University Press, 1993), p. 128.

8. Postel, *Pillar of Sand,* p. 32.

9. Ibid., p. 30.

10. William R. Jobin, *Dams and Disease: Ecological Design and Health Impacts of Large Dams, Canals, and Irrigation Systems* (New York: Routledge, 1999), p. 295.

11. Fred Pearce, "High and Dry in Aswan," *New Scientist* 142 (May 1994).

12. Ibid.

13. Craig Childs, "Phoenix Falling?" *High Country News,* 16 April 2007. This is a fine general introduction to the Hohokam story.

14. Ibid.

15. John Andrews and Todd Bostwick, *Desert Farmers at the River's Edge: The Hohokam and Pueblo Grande* (City of Phoenix, 2000).

16. David M. Meko, Connie A. Woodhouse, C. A. Baisan, T. Knight, J. Lukas, M. Hughes, and M. Salzer, "Medieval Drought in the Upper Colorado River Basin," *Geophysical Research Letters* 34 (2007).

17. George Thomas, *The Development of Institutions under Irrigation* (New York: Macmillan, 1920), p. 27.

CHAPTER FOUR

1. John A. Widtsoe, *Success on Irrigation Projects* (New York: John Wiley, 1928), p. 138.

2. Donald Worster, *A River Running West: The Life of John Wesley Powell* (New York: Oxford University Press, 2002), p. 339.

3. Wallace Earle Stegner, *Beyond the Hundredth Meridian: John Wesley Powell and the Second Opening of the West* (New York: Penguin Books, 1992), p. 45.

4. John Wesley Powell, *Report on the Lands of the Arid Region of the United States, with a More Detailed Account of the Lands of Utah* (facsimile of the 1879 ed., Boston, Mass.: Harvard Common Press, 1983).

5. Ibid., p. xi (introduction by T. H. Watkins).

6. Worster, *A River Running West*, p. 355.

7. Marc Reisner, *Cadillac Desert: The American West and Its Disappearing Water* (New York: Penguin Books, 1987). p. 47.

8. Stegner, *Beyond the Hundredth Meridian*, p. 212.

9. Ibid., p. 214.

10. Powell, *Report on the Lands of the Arid Region*, p. 27.

11. Ibid., p. 9.

12. Utah Agricultural Statistics Service, "2003 Farm and Ranch Irrigation Survey Results Released," http://www.nass.usda.gov.

13. *Congressional Record*, 11 July 1882, pp. 5923–5930.

14. John Wesley Powell, "The Work of the U.S. Geological Survey," *Science* 21, no. 519 (1893).

15. Worster, *A River Running West*, p. 529.

16. Quotations are from John Wesley Powell, "Official Proceedings of the International Irrigation Congress" (Los Angeles, 1893), pp. 106–12.

17. Public Broadcasting Service, "People and Events: Water Fights," http://www.pbs.org.

18. Stegner, *Beyond the Hundredth Meridian*, p. 338.

19. John Wesley Powell and William Eno DeBuys, *Seeing Things Whole: The Essential John Wesley Powell* (Washington, D.C.: Island Press/Shearwater Books, 2001), p. 231.

20. John Wesley Powell, "The Lesson of Conemaugh," *North American Review* 149 (1889): 150–56.

21. Theodore Roosevelt, "First Annual Message to Congress," *New York Times*, 4 December 1901.

22. Michael C. Robinson, *Water for the West: The Bureau of Reclamation, 1902–1977* (Chicago: Public Works Historical Society, 1979), p. 165.

23. Powell, *Report on the Lands of the Arid Region*, p. 42.

24. Worster, *A River Running West*, p. 357.

CHAPTER FIVE

1. Donald Pisani, *Water and American Government* (Berkeley: University of California Press, 2002), p. 30.

2. Shelly C. Dudley, "The First Five: A Brief History of the Salt River Project," http://www.waterhistory.org/histories/reclamation/saltriver.

3. Donald Worster, *Rivers of Empire: Water, Aridity, and the Growth of the American West* (New York: Pantheon Books, 1985), p. 172.

4. *Reclamation Act/Newlands Act of 1902*, 43 U.S. Code 371.

5. Theodore Roosevelt, "Roosevelt Describes Work of Irrigation," *New York Times*, 4 September 1906.

6. Quoted in Worster, *Rivers of Empire*, p. 173.

7. Dorothy Lampen, *Economic and Social Aspects of Federal Reclamation* (Manchester, N.H.: Ayer Publishing, 1979), p. 60.

8. Michael C. Robinson, *Water for the West: The Bureau of Reclamation, 1902–1977* (Chicago: Public Works Historical Society, 1979), p. 38.

9. Ibid., p. 37.

10. National Research Council, *Managing Construction and Infrastructure in the 21st Century Bureau of Reclamation* (Washington, D.C.: National Academies Press, 2006).

11. Robinson, *Water for the West*, p. 38.

12. Committee of Special Advisors on Reclamation 1923–24, *Federal Reclamation by Irrigation*, 68th Cong., 1st sess., Senate Doc. 92 (Serial Set 8238), 1924.

13. Worster, *Rivers of Empire*, p. 179.

14. Ibid., p. 179.

15. Frederick Newell, *Irrigation Management* (New York: D. Appleton and Co., 1916), pp. 2, 31.

16. Committee of Special Advisors on Reclamation 1923–24, *Federal Reclamation by Irrigation*, p. xi.

17. Loomis Havemeyer, ed., et al., *Conservation of Our Natural Resources, Based on [Charles Richard] Van Hise's "The Conservation of Natural Resources in the United States"* (New York: Macmillan, 1930), p. 154.

18. Ibid., p. 163.

19. John A. Widtsoe, *Success on Irrigation Projects* (New York: John Wiley, 1928), p. 138.

20. Quoted in Norris Hundley, Jr., "The West against Itself," in *New Courses for the Colorado River: Major Issues for the Next Century*, ed. Gary D. Weatherford and F. Lee Brown (Albuquerque: University of New Mexico Press, 1986), pp. 10–11.

CHAPTER SIX

1. Imperial Irrigation District, "General History" (2006), http://www.iid.com.

2. Robert L. Sperry, "When the Imperial Valley Fought for Its Life," *The Journal of San Diego History* 21 (Winter 1975).

3. William E. Smythe, "An International Wedding: The Tale of a Trip on the Borders of Two Republics," *Sunset* 5, no. 3 (1900): 293–94.

4. Norris Hundley, Jr., *The Great Thirst: Californians and Water: A History*, rev. ed. (Berkeley: University of California Press, 2001), p. 208.

5. Quoted in Hundley, "The West against Itself," in *New Courses for the Colorado River: Major Issues for the Next Century*, ed. Gary D. Weatherford and F. Lee Brown (Albuquerque: University of New Mexico Press, 1986), pp. 12–13.

6. *The Oxford English Dictionary*, 2nd ed. (New York: Oxford University Press, 1989).

7. Gary Topping, *Glen Canyon and the San Juan Country* (Moscow: University of Idaho Press, 1997), pp. 336–38.

8. John Andreas Widtsoe, *In a Sunlit Land: The Autobiography of John A. Widtsoe* (Salt Lake City: Deseret News Press, 1952), p. 180.

9. Ralph L. Carr, "Delph Carpenter: Father of Colorado River Treaties" (text of 1943 Salute, Colorado State University, 1991).

10. Ibid.

11. R. Eric Kuhn, "Future Scenarios for the Colorado River" (paper presented at the American Bar Association, San Diego, February 24–25, 2005).

12. Colorado River Commission, "Colorado River Compact" (1922), http://www.colorado.edu/resources/colorado_river/law_policy.html. The Colorado River Compact refers to "Lee Ferry" and defines it as a point one mile below the mouth of the Paria River. Lee's Ferry, the home of John D. Lee, the first settler at the site, is located at the mouth of the Paria. The U.S. Board on Geographic Names disregards apostrophes and uses Lees Ferry. I use Lee's Ferry throughout but will give the clipped spelling "Lee Ferry" where it appears in quotations.

13. Ibid.

14. Quoted in W. B. Langbein, "L'affaire Larue," *U.S. Geological Survey Water Resources Division Bulletin,* WRD Historical Note I (1975), p. 44.

15. Ibid., p. 46.

16. Hundley, *The Great Thirst,* p. 214.

CHAPTER SEVEN

1. Quoted in Public Broadcasting Service, "The Men of Six Companies" (1999), http://www.pbs.org.

2. Dennis McBride, "Frank Crowe: Crowning Achievement," in *The First 100: Portraits of the Men and Women Who Shaped Las Vegas,* ed. A. D. Hopkins and K. J. Evans (Las Vegas: Huntington Press, 1999), pp. 95–98.

3. Ibid.

4. Public Broadcasting Service, "The Men of Six Companies."

5. Bureau of Reclamation, "Concrete" (2005), http://www.usbr.gov/lc/hooverdam/History/essays/concrete.html.

6. Bureau of Reclamation, "Tunnels" (2004), http://www.usbr.gov/lc/hooverdam/History/essays/tunnels.html.

7. Bureau of Reclamation, "High Scalers" (2004), http://www.usbr.gov/lc/hooverdam/History/essays/hscaler.html.

8. Ibid.

9. Marc Reisner, *Cadillac Desert: The American West and Its Disappearing Water* (New York: Penguin Books, 1987), p. 134.

10. "Owyhee" is an old spelling of Hawaii, originating with Captain James Cook. The Owyhee River was named for three Hawaiian trappers killed by Indians.

11. Bureau of Reclamation, "Concrete."

12. Bureau of Reclamation, "Artwork" (2001), http://www.usbr.gov/lc/hooverdam/History/essays/artwork.html.

13. Quoted in Joseph E Stevens, *Hoover Dam: An American Adventure* (Norman: University of Oklahoma, 1990), p. 248.

14. Bureau of Reclamation, "Fatalities at Hoover Dam," (2004), http://www.usbr.gov/lc/hooverdam/History/essays/fatal.html.

15. Quoted in Dennis McBride, "Desperate Times: The Building of Hoover Dam" (2003), http://www.bbc.co.uk.

16. John Tauranac, *Empire State Building: The Making of a Landmark* (New York: St. Martin's Griffin, 1997), p. 221.

17. "Glen Canyon Dam" (2006), http://www.powellguide.com.

18. Public Broadcasting Service, "People and Events: Frank Crowe," http://www.pbs.org/wgbh/amex/hoover/peopleevents/pandeAMEX87.html.

19. McBride, "Frank Crowe."

20. Quoted in "The Hoover Dam: Lonely Lands Made Fruitful," http://xroads.virginia.edu.

21. Quoted in James M. Smith, *The Republic of Letters: The Correspondence between Thomas Jefferson and James Madison 1776–1826* (New York: W.W. Norton, 1995), p. 334.

22. *Congressional Record*, "Remarks of Francis Newlands," 13 June 1902, p. 6734.

23. Norris Hundley, Jr., *The Great Thirst: Californians and Water: A History*, rev. ed. (Berkeley: University of California Press, 2001), p. 225.

24. *Boulder Canyon Project Act*, 43 U.S. Code 618.

25. Hundley, *The Great Thirst*, p. 223.

26. Philip L. Fradkin, *A River No More: The Colorado River and the West* (Berkeley: University of California Press, 1996), p. 280.

27. Quoted in ibid, p. 276.

28. Donald Worster, *Rivers of Empire: Water, Aridity, and the Growth of the American West* (New York: Pantheon Books, 1985), p. 212.

29. *Bryant v. Yellen* (1980), 447 U.S. Code 352.

30. Philip Martin, "Imperial Valley: Agriculture and Farm Labor" (2001), http://migration.ucdavis.edu.

31. University of California at Davis, "Irrigation Management and Water Quality" (2006), http://tmdl.ucdavis.edu.

32. National Resources Defense Council, "Alfalfa: The Thirstiest Crop" (2001), http://www.nrdc.org.

33. University of California at Davis, "Imperial County Agriculture" (2006), http://commserv.ucdavis.edu/CEImperial/overview.htm.

CHAPTER EIGHT

1. Michael C. Robinson, *Water for the West: The Bureau of Reclamation, 1902–1977* (Chicago: Public Works Historical Society, 1979), pp. 55–56.

2. Ibid., p. 56.

3. Infoplease, "Life Expectancy by Age, 1850–2004" (2006), http://www.infoplease.com/ipa/A0005140.html.

4. Norris Hundley, Jr., *The Great Thirst: Californians and Water: A History,* rev. ed. (Berkeley: University of California Press, 2001).

5. Bureau of Reclamation, *The Colorado River: A Natural Menace Becomes a National Resource* (Washington, D.C.: Department of Interior, 1946); known as the "Blue Book."

6. Marc Reisner, *Cadillac Desert: The American West and Its Disappearing Water* (New York: Penguin Books, 1987), p. 143.

7. Mark W. T. Harvey, *A Symbol of Wilderness: Echo Park and the American Conservation Movement* (Albuquerque: University of New Mexico Press, 1994).

8. National Park Service, "Public Use Statistics Office," http://www.nature.nps.gov/stats.

9. Harvey, *Symbol of Wilderness,* p. 159.

10. William J. Clinton, "Remarks Announcing the Establishment of Grand Staircase–Escalante National Monument," *Weekly Compilation of Presidential Documents* 32, 1785 (23 September 1996).

11. Joe Baird, "Utah: Trim President's Monument Powers," *Salt Lake Tribune,* 29 January 2005.

12. Bureau of Reclamation, *Colorado River: A Natural Menace,* p. 18.

13. Ibid., p. 218.

14. Quoted in Reisner, *Cadillac Desert,* p. 146.

15. Bureau of Reclamation, *Colorado River: A Natural Menace,* p. 3.

16. Upper Colorado River Basin Commission, "Upper Colorado River Basin Compact" (1948), http://www.usbr.gov/lc/region/pao/pdfiles/ucbsnact.pdf.

17. Quoted in Robinson, *Water for the West,* p. 79.

18. Ibid., p. 80.

19. Jim Stiles, "Let's Drain Lake Powell," *Canyon Country Zephyr* (April/May 1997).

20. David Brower, "Oral Interview" (Bancroft Library, University of California at Berkeley, 1979), p. III.

21. Eric Brazil, "David Brower Dies at 88," *San Francisco Examiner,* 6 November 2000.

CHAPTER NINE

1. *National Park Service Act,* 16 U.S. Code 221 et seq.

2. *The Oxford English Dictionary,* 2nd ed. (New York: Oxford University Press, 1989).

3. Ibid.

4. Quoted in Mark W. T. Harvey, *A Symbol of Wilderness: Echo Park and the American Conservation Movement* (Albuquerque: University of New Mexico Press, 1994), p. 32.

5. Ibid., p. 33.

6. Split Mountain is several miles downstream of Echo Park. The Bureau of Reclamation soon dropped plans for a reservoir at the site.

7. Bureau of Reclamation, *The Colorado River: A Natural Menace Becomes a National Resource* (Washington, D.C.: Department of Interior, 1946), pp. 244–45.

8. House Committee on Interior and Insular Affairs, Subcommittee on Irrigation and Reclamation, *Colorado River Storage Project: Hearings on H.R. 4449, 4443, and 4463*, 83rd Cong., 2nd sess. (Washington, D.C.: U.S. Government Printing Office, 1954), p. 595.

9. National Park Service, "A Survey of the Recreational Resources of the Colorado River Basin" (Washington, D.C.: National Park Service, 1950), pp. 197–98.

10. Harvey, *Symbol of Wilderness*, p. 79.

11. Department of the Interior, *Proceedings before the United States Department of the Interior: Hearing on Dinosaur National Monument, Echo Park, and Split Mountain Dams* (Washington, D.C., 1950), p. 446.

12. Jon M. Cosco, *Echo Park: Struggle for Preservation* (Boulder, Colo.: Johnson Books, 1995), p. 30.

13. Department of the Interior, *Proceedings before the United States Department of the Interior: Hearing on Dinosaur National Monument, Echo Park, and Split Mountain Dams*, p. 8.

14. Ibid., pp. 185–86.

15. Ibid., p. 299.

16. Bernard DeVoto, "Shall We Let Them Ruin Our National Parks?" *Saturday Evening Post*, 22 July 1950, pp. 17–19, 42, 44, 46, 48.

17. Department of the Interior, *Proceedings before the United States Department of the Interior: Hearing on Dinosaur National Monument, Echo Park, and Split Mountain Dams*, p. 375.

18. Ibid., p. 377.

19. Ulysses S. Grant III, "Supplementary Report by General Grant on the Dinosaur Controversy," *Planning and Civic Comment* 17 (1951): 4.

20. "Decision by Oscar Chapman, Secretary of the Interior, Regarding the Dinosaur National Monument Controversy" (1950), in House Committee on Interior and Insular Affairs, *Colorado River Storage Project: Hearings on H.R. 4449, 4443*, and *4463*, p. 446.

21. Harvey, *Symbol of Wilderness*, pp. 90–91.

22. Jerry N. Hess, "Oral History Interviews with Oscar L. Chapman" (1972), http://www.trumanlibrary.org.

23. Bernard DeVoto, "The West against Itself," *Harpers*, January 1947.

24. DeVoto, "Shall We Let Them Ruin Our National Parks?"

25. Ibid., p. 42.

26. Phillip L. Fradkin, *A River No More: The Colorado River and the West* (Berkeley: University of California Press, 1996), p. 191.

27. Bernard DeVoto, "Letter to Editor," *Denver Post,* 1 August 1950 (DeVoto's italics); Cosco, *Echo Park,* p. 46.

28. Quoted in "Let's Be Fair, Mike," *Nature Magazine* 44 (October 1951): 425.

29. Harvey, *Symbol of Wilderness,* p. 100.

30. Hess, "Oral History Interviews with Oscar L. Chapman."

31. Ibid.

32. Newton Drury, "Oral Interview" (Bancroft Library, University of California at Berkeley, 1972), p. 523.

33. Tom Turner, *Sierra Club: 100 Years of Protecting Nature* (New York: H.N. Abrams in association with the Sierra Club, 1991), p. 48.

34. *Grand Canyon Protection Act,* Public Law 102–575, 106 Stat. 4669.

35. Sierra Club, "Minutes of a Special Meeting of the Board of Directors," *Sierra Club Magazine* 34, no. 11 (1949): 3–4.

36. Turner, *Sierra Club: 100 Years of Protecting Nature*, pp. 131–32.

CHAPTER TEN

1. Quoted in W. L. Rusho, "Bumpy Road for Glen Canyon Dam" (paper presented at the symposium on the history of the Bureau of Reclamation, University of Nevada at Las Vegas, 2002), p. 20.

2. Ken Verdoia, "Interview: Floyd Dominy," http://www.kued.org/productions/glencanyon/interviews/dominy.html.

3. Quoted in Jon M. Cosco, *Echo Park: Struggle for Preservation* (Boulder, Colo.: Johnson Books, 1995), p. 60.

4. House Committee on Interior and Insular Affairs, Subcommittee on Irrigation and Reclamation, *Colorado River Storage Project: Hearings on H.R. 4449, 4443, and 4463,* 83rd Cong., 2nd sess. (Washington, D.C.: U.S. Government Printing Office, 1954), pp. 833–34.

5. Quoted in Stephen C. Sturgeon, *The Politics of Western Water: The Congressional Career of Wayne Aspinall* (Tucson: University of Arizona Press, 2002), p. 47.

6. Dennis Farney, "Ruler of the Land," *Wall Street Journal,* 21 January 1972.

7. University of Virginia, "Secretary of the Interior: Douglas J. McKay (1953–1956)" (2003), http://www.americanpresident.org/history.

8. Ralph Tudor, "I'm Glad I Came to Washington," *Saturday Evening Post,* 27 November 1954.

9. House Committee on Interior and Insular Affairs, Subcommittee on Irrigation and Reclamation, *Colorado River Storage Project: Hearings on H.R. 4449, 4443, and 4463,* pp. 18, 22, 792.

10. Quoted in Jon M. Cosco, *Echo Park: Struggle for Preservation* (Boulder, Colo.: Johnson Books, 1995), p. 71.

11. Quoted in ibid., p. 127.

12. Mark W. T. Harvey, *A Symbol of Wilderness: Echo Park and the American Conservation Movement* (Albuquerque: University of New Mexico Press, 1994), p. 191.

13. House Committee on Interior and Insular Affairs, Subcommittee on Irrigation and Reclamation, *Colorado River Storage Project: Hearings on H.R. 4449, 4443, and 4463*, p. 825.

14. Ibid., pp. 824–25.

15. Ibid., p. 825.

16. David Brower, "Oral Interview" (Bancroft Library, University of California at Berkeley, 1979).

17. Senate Committee on Interior and Insular Affairs, Subcommittee on Irrigation and Reclamation, *Colorado River Storage Project: Hearings*, 83rd Cong., 2nd sess., June 28–30, July 1–3, 1954, on S. 1555, p. 48.

18. Brower, "Oral Interview," p. 117.

19. House Committee on Interior and Insular Affairs, Subcommittee on Irrigation and Reclamation, *Colorado River Storage Project: Hearings on H.R. 4449, 4443, and 4463*, p. 505.

20. Senate Committee on Interior and Insular Affairs, *Colorado River Storage Project: Hearings*, 83rd Cong., 2nd sess., June 28–30, July 1–3, 1954, pp. 5–6.

21. Senate Committee on Interior and Insular Affairs, Subcommittee on Irrigation and Reclamation, *Colorado River Storage Project: Hearings*, 84th Cong., 1st sess., February 28, March 1, 2, 3, 4, and 5, 1955, on S. 500, p. 639.

22. *Congressional Record*, "Remarks of Wayne Aspinall," 16 August 1954, pp. 14698–14700.

23. Quoted in Cosco, *Echo Park*, p. 75.

24. Wallace Earle Stegner, *This Is Dinosaur: Echo Park Country and Its Magic Rivers* (New York: Knopf, 1955), insert.

25. Quoted in Sturgeon, *Politics of Western Water*, p. 38.

26. Brower, "Oral Interview."

27. Leslie A. Miller, "Dollars into Dust," *Reader's Digest*, May 1955, p. 112.

28. "Bananas on Pike's Peak?" *New York Times*, 24 April 1955.

29. Quoted in Sturgeon, *Politics of Western Water*, p. 47.

30. Cosco, *Echo Park*, p. 88.

31. *Colorado River Storage Project*, 43 U.S. Code 620.

32. Robert T. Hartmann, "Lawmaker Hurls Rock into Fight against Dam," *Los Angeles Times*, 29 January 1956.

33. *Congressional Record*, "Remarks of Stewart Udall," 1 March 1956, p. 3736.

34. Russell Martin, *A Story That Stands Like a Dam: Glen Canyon and the Struggle for the Soul of the West* (Salt Lake City: University of Utah Press, 1999), see p. 72 for an amusing account of these experiments.

35. Howard Zahniser, "How Much Wilderness Can We Afford to Lose?" (address at the Second Wilderness Conference, Oakland, Calif., 30 March 1951); quoted in Cosco, *Echo Park,* pp. 94–95.

36. House Committee on Interior and Insular Affairs, Subcommittee on Irrigation and Reclamation, *Colorado River Storage Project: Hearings on H.R. 4449, 4443, and 4463,* p. 827.

37. Ken Verdoia, "Interview with David Brower," http://www.kued.org/productions/glencanyon/interviews/brower.html.

38. Wallace Earle Stegner, *This Is Dinosaur: Echo Park Country and Its Magic Rivers* (Boulder, Colo.: Roberts Rinehart, 1985), p. vii.

39. Brower, "Oral Interview," p. 118.

CHAPTER ELEVEN

1. Ken Verdoia, "Interview: Floyd Dominy," http://www.kued.org/productions/glencanyon/interviews/dominy.html.

2. National Park Service, "Rainbow Bridge: Administrative History" (2003), http://www.nps.gov/archive/rabr/adhi/adhi6a.htm.

3. Verdoia, "Interview: Floyd Dominy."

4. Ibid.

5. Ibid.

6. National Park Service, "Rainbow Bridge."

7. Quoted in Russell Martin, *A Story That Stands Like a Dam: Glen Canyon and the Struggle for the Soul of the West* (Salt Lake City: University of Utah Press, 1999), p. 220.

8. National Park Service, "Rainbow Bridge."

9. Martin, *A Story That Stands Like a Dam,* pp. 305–306.

10. David Brower, "Rainbow Promise Breaking; New Threat at Echo Park," *Sierra Club Bulletin* (April–May 1960), p. 20.

11. David Brower, "Oral Interview" (Bancroft Library, University of California at Berkeley, 1979), p. 90, app. A.

12. David Brower, "Please Keep Those Glen Canyon Tunnels Open until Rainbow Bridge Protection Is Certain," *Sierra Club Bulletin* 47 (March–April 1962): 2–3 (Brower's italics).

13. National Park Service, "Rainbow Bridge."

14. Martin, *A Story That Stands Like a Dam,* pp. 304–305; National Park Service, "Rainbow Bridge."

15. Verdoia, "Interview: Floyd Dominy."

16. Eliot Porter and David Ross Brower, *The Place No One Knew: Glen Canyon on the Colorado,* commemorative ed. (Salt Lake City: Gibbs Smith, 2000), p. 13.

17. Eliot Porter and David Brower, *The Place No One Knew: Glen Canyon on the Colorado,* abridged ed. (San Francisco: Sierra Club, 1968), foreword, p. 5.

18. Bureau of Reclamation, *Lake Powell, Jewel of the Colorado* (Washington, D.C.: U.S. Government Printing Office, 1965).

19. David Wegner, personal communication.

20. Porter and Brower, *The Place No One Knew* (commemorative ed., 2000), p. 168.

21. Quoted in Martin, *A Story That Stands Like a Dam*, p. 246.

22. Wallace Stegner, "Glen Canyon Submersis," in *The Glen Canyon Reader*, ed. Mathew Barrett Gross (Tucson: University of Arizona Press, 2003), p. 142.

23. Bureau of Reclamation, *The Colorado River: A Natural Menace Becomes a National Resource* (Washington, D.C.: Department of Interior, 1946).

24. Marc Reisner, *Cadillac Desert: The American West and Its Disappearing Water* (New York: Penguin Books, 1987), p. 283.

25. National Park Service, "Public Use Statistics Office," http://www2 .nature.nps.gov/stats.

26. Quoted in Martin, *A Story That Stands Like a Dam*, p. 270.

27. Quoted in ibid., p. 271.

28. Tom Turner, "The Grand Undammed—Proposals to Dam Grand Canyon," *Sierra Club,* July–August 1992.

29. Ibid.

30. Martin, *A Story That Stands Like a Dam*, p. 273.

31. Turner, "The Grand Undammed."

32. House Committee on Interior and Insular Affairs, Subcommittee on Irrigation and Reclamation, *Hearings, Colorado River Basin Project* (Washington, D.C.: U.S. Government Printing Office, 1967), pp. 458–59.

33. Turner, "The Grand Undammed."

34. Montgomery Brower, "David Brower: President of Friends of the Earth," *People Weekly* 33 (1990): 103–6.

35. *Canyon County Zephyr*, "An Interview with Dr. Rich Ingebretsen" (August–September 2000).

36. Quoted in Reisner, *Cadillac Desert*, p. 298.

37. L. B. Leopold, "Oral Interview" (Bancroft Library, University of California at Berkeley, 1993), p. 185.

38. Ibid., p. 183.

39. Martin, *A Story That Stands Like a Dam*, pp. 281–82.

40. Leopold, "Oral Interview," p. 187.

41. Lew Stieger, "David Brower," *Boatman's Quarterly Review,* Summer 1997.

42. Quoted in Roderick Nash, *Grand Canyon of the Living Colorado* (San Francisco: Sierra Club, 1970), p. 105.

43. Reisner, *Cadillac Desert*, p. 302.

44. *Colorado River Basin Project,* 43 U.S. Code 1501–1506.

45. Reisner, *Cadillac Desert*, p. 260.

46. T. J. Wolf, "Mr.Dominy, Are You a Hero or a Villain?" *High Country News,* 26 October 1998.

47. Martin, *A Story That Stands Like a Dam,* p. 292.

48. Ibid., pp. 292–93.

49. John A. McPhee, *Encounters with the Archdruid* (New York: Farrar Straus and Giroux, 1971).

50. David Wegner, personal communication.

51. Bernard DeVoto, "Shall We Let Them Ruin Our National Parks?" *Saturday Evening Post,* 22 July 1950, p. 46.

52. Richard J. Ingebretsen, "Floyd Dominy, the Napkin, and Me" (1999), http://www.glencanyon.org.

CHAPTER TWELVE

1. Richard L. Berkman and W. Kip Viscusi, *Damming the West*, Ralph Nader's Study Group Report on the Bureau of Reclamation (New York: Grossman Publishers, 1973), p. 6.

2. *Congressional Record*, "Construction of the Colorado River Storage Project," 18 April 1955, pp. 4577, 4576.

3. Ibid., p. 4576.

4. Ibid., p. 4578.

5. Ibid., p. 4641.

6. Ibid.

7. Ibid., p. 4638.

8. John A. Widtsoe, *Success on Irrigation Projects* (New York: John Wiley, 1928), p. 140.

9. Berkman and Viscusi, *Damming the West*, p. 79.

10. Ibid., p. 79.

11. Quoted in Leslie A. Miller, "The Battle That Squanders Billions," *Saturday Evening Post* (14 May 1949), pp. 30–31, 160–62.

12. Ibid.

13. Central Arizona Project, "About Cap-Faq" (1997), http://www.capaz.com.

14. Miller, "Battle That Squanders Billions."

15. Leslie A. Miller, "Dollars into Dust," *Reader's Digest* (May 1955), pp. 112–13.

16. Marc Reisner, *Cadillac Desert: The American West and Its Disappearing Water* (New York: Penguin Books, 1987), p. 302.

17. Quoted in ibid.

18. Arthur Jones, "Where Water Is Power, Pork Is Swimming," *National Catholic Reporter*, 26 December 1997.

19. General Accounting Office, "Animas–La Plata Project," GAO/RCED-96–1 (Washington, D.C.: U.S. Government Printing Office, 1995), p. 11.

20. Michael Satchell, "The Last Water Fight," *U.S. News & World Report,* 23 October 1995.

21. Gail Binkly, "A-LP [Animas–La Plata] Gets Federal A-Ok," *High Country News,* 27 August 2001.

22. Gail Binkly, "Dam's Price Tag Skyrockets," *High Country News,* 10 May 2004.

23. Miller, "Battle That Squanders Billions."

24. Reisner, *Cadillac Desert,* p. 176.

25. Quoted in John P. Gregg, "Mass. Appeal," *New York Times,* 25 August 2005.

26. L. B. Leopold, *Probability Analysis Applied to a Water-Supply Problem* (Reston, Va.: U.S. Geological Survey, 1958), p. 14.

27. Ibid., p. 15.

28. Leopold, "Oral Interview," p. 184.

29. *Congressional Record,* "Construction of the Colorado River Storage Project," p. 4574.

30. Ken Verdoia, "Interview: Floyd Dominy," http://www.kued.org/productions/glencanyon/interviews/dominy.html.

31. Glen Canyon Institute, "Frequently Asked Questions" (2006), http://www.glencanyon.org.

32. Bruce Berger, *There Was a River* (Tucson: University of Arizona Press, 1994), pp. 45–47 (Berger's italics).

33. *Congressional Record.* "Construction of the Colorado River Storage Project," p. 4574.

CHAPTER THIRTEEN

1. Michelle Nijhuis, "Written in the Rings," *High Country News,* 24 January 2005.

2. Charles Stockton and Gordon Jacoby, Jr., "Long-Term Surface-Water Supply and Streamflow Trends in the Upper Colorado River Basin," *Lake Powell Research Project Bulletin,* no. 18 (1976): 1–73.

3. Hugo G. Hidalgo, Thomas C. Piechota, and John A. Dracup, "Alternative Principal Components Regression Procedures for Dendrohydrologic Reconstructions," *Water Resources Research* 36 (2000): 3241–49.

4. Connie A. Woodhouse, Stephen T. Gray, and David M. Meko, "Updated Streamflow Reconstructions for the Upper Colorado River Basin," *Water Resources Research* 42 (2006).

5. David M. Meko, Connie A. Woodhouse, et al., "Medieval Drought in the Upper Colorado River Basin," *Geophysical Research Letters* 34 (2007).

6. Bureau of Reclamation, *Final Environmental Impact Statement, Colorado River Interim Surplus Criteria,* vol. 2 (2000), attachment K.

7. Charles Dickens, *David Copperfield* (New York: Random House, 2000), p. 161.

8. Tim P. Barnett, David W. Pierce, Hugo G. Hidalgo, Celine Bonfils, Benjamin D. Santer, Tapash Das, Govindasamy Bala, Andrew W. Wood, Toru Nozawa, Arthur A. Mirin, Daniel R. Cayan, and Michael D. Dettinger, "Human-Induced Changes in the Hydrology of the Western United States," *Science* 319, 5866 (2008).

9. IPCC, 2001, *Climate Change 2001: Synthesis Report. Contribution of Working Groups I, II, and III to the Third Assessment Report of the Intergovernmental Panel on Climate Change,* ed. R. T. Watson, Daniel L. Albritton et al. (Cambridge and New York: Cambridge University Press, 2001).

10. IPCC, 2007, "Summary for Policymakers," in *Climate Change 2007: The Physical Science Basis. Contribution of Working Group I to the Fourth Assessment Report of the Intergovernmental Panel on Climate Change,* ed. Susan Solomon, Dahe Qin et al. (Cambridge and New York: Cambridge University Press, 2007), p. 8.

11. National Oceanic and Atmospheric Administration, "U.S. Has Its Second-Hottest July on Record," 7 August 2006, http://www.noaanews.noaa.gov/stories2006/s2677.htm.

12. National Oceanic and Atmospheric Administration, "2006 Annual Climate Review, U.S. Summary" (2007), http://www.ncdc.noaa.gov.

13. National Research Council, *Surface Temperature Reconstructions for the Last 2,000 Years* (Washington, D.C.: National Academies Press, 2006).

14. Andrew Revkin, "Connecting the Global Warming Dots," *New York Times,* 14 January 2007.

15. Sharon Begley and Andrew Murr, "Which of These Is Not Causing Global Warming Today?" *Newsweek,* 24 June 2007.

16. Naomi Oreskes, "The Scientific Consensus on Climate Change," *Science* 306 (2004): 1686.

17. IPCC, 2007, "Summary for Policymakers."

18. Richard A. Kerr, "Pushing the Scary Side of Global Warming," *Science* 316 (2007): 1412–15.

19. IPCC, 2007, "Summary for Policymakers," p. 9.

20. K. M. Walter, S. A. Zimov, J. P. Chanton, D. Verbyla, and F. S. Chapin, "Methane Bubbling from Siberian Thaw Lakes as a Positive Feedback to Climate Warming," *Nature* 4443 (2006): 71–75.

21. Doug Struck, "'Rapid Warming' Spreads Havoc in Canada's Forests," *Washington Post,* 1 March 2006.

22. Stefan Rahmstorf, Anny Cazenave, John A. Church, James E. Hansen, Ralph F. Keeling, David E. Parker, and Richard C. J. Somerville, "Recent Climate Observations Compared to Projections," *Science* 316 (2007).

23. United Nations, *Adaptation and Vulnerability to Climate Change: The Role of the Finance Sector,* United Nations Environment Programme Finance Initiative (Geneva, 2006).

24. Ibid., p. 4.

25. IPCC, 2007, "Summary for Policymakers," p. 12.

26. M. Baettig, M. Wild, and D. M. Imboden, "A Climate Change Index: Where Climate Change May Be Most Prominent in the 21st Century," *Geophysical Research Letters* 34 (2007).

27. Rocky Mountain Climate Organization, *Less Snow, Less Water: Climate Disruption in the West* (Louisville, Colo., 2005).

28. Steven W. Running, "Is Global Warming Causing More, Larger Wildfires?" *Science* 313 (2006): 927–28.

29. Robert F. Service, "As the West Goes Dry," *Science* 303 (2004).

30. Iris T. Stewart, D. Cayan, and M. D. Dettinger, "Changes in Snowmelt Runoff Timing in Western North America under a 'Business as Usual' Climate Change Scenario," *Climatic Change* 62 (2004): 217–32.

31. IPCC, 2007, "Summary for Policymakers," p. 11.

32. Rocky Mountain Climate Organization, *Less Snow, Less Water.*

CHAPTER FOURTEEN

1. Niklas S. Christensen, Andrew W. Wood, Nathalie Voisin, Dennis P. Lettenmaier, and Richard N. Palmer, "The Effects of Climate Change on the Hydrology and Water Resources of the Colorado River Basin," *Climatic Change* 62 (2004): 342.

2. Christensen et al., "Effects of Climate Change on the Hydrology and Water Resources of the Colorado River Basin."

3. Niklas S. Christensen and Dennis P. Lettenmaier, "A Multimodel Ensemble Approach to Assessment of Climate Change Impacts on the Hydrology and Water Resources of the Colorado River Basin," *Hydrology and Earth System Sciences Discussions* 3 (2006): 3727–3770.

4. P. C. D. Milly, K. A. Dunne, and A. V. Vecchia, "Global Pattern of Trends in Streamflow Availability in a Changing Climate," *Nature* 438 (2005): 347–50.

5. Martin Hoerling and Jon Eischeid, "Past Peak Water in the Southwest," *Southwest Hydrology* (January–February 2007), pp. 18–19, 35.

6. R. Seager, M. Ting, I. Held, Y. Kushnir, J. Lu, G. Vecchi, H. Huang, N. Harnik, A. Leetmaa, N. Lau, C. Li, J. Velez, and N. Naik, "Model Projections of an Imminent Transition to a More Arid Climate in Southwestern North America," *Science* 316 (2007): 1181–84.

7. IPCC, 2007, "Summary for Policymakers," in *Climate Change 2007: Impacts, Adaptation and Vulnerability. Contribution of Working Group II to the Fourth Assessment Report of the Intergovernmental Panel on Climate Change,* ed. W. Neil Adger et al. (Geneva, Switzerland: WMO, 2007).

8. Bureau of Reclamation and U.S. Fish and Wildlife Service, *Lower Colorado River Multi-Species Conservation Program: Final Programmatic Environmental Impact Statement, Environmental Impact Report,* ed. Department of the

Interior, 5 vols. (Washington, D.C.: Department of the Interior, 2004); see final appendices to vols. 1–4, sect. J.7.1.

9. Ibid., "Comments and Responses," sect. 3, p. 8.

10. National Academy of Sciences, *Colorado River Basin Water Management: Evaluating and Adjusting to Hydroclimatic Variability* (Washington, D.C.: National Academies Press, 2007).

11. Ibid., p. 6.

12. Quotes in Bureau of Reclamation, *Final Environmental Impact Statement, Colorado River Interim Guidelines for Lower Basin Shortages and Coordinated Operations for Lake Powell and Lake Mead* (Washington, D.C.: Department of the Interior, 2007), vol. 4, pp. 9, 14.

13. Elizabeth Rosenthal, "U.N. Chief Seeks More Climate Change Leadership," *New York Times,* 18 November 2007.

14. White House, "Press Briefing via Conference Call by Senior Administration Officials on IPCC Report," 16 November 2007.

15. Arizona State University, "Watersim Model," http://watersim.asu.edu.

16. Living Rivers, "Colorado River Open Source Simulator," http://onthe-colorado.org/cross.cfm.

CHAPTER FIFTEEN

1. United Nations Development Programme, Human Development Report 2006, *Beyond Scarcity: Power, Poverty and the Global Water Crisis* (New York: Palgrave Macmillan, 2006), p. 135.

2. Ibid., p. 135.

3. Ibid., pp. 137, 135.

4. C. M. Wong, C. E. Williams, J. Pittock, U. Collier, and P. Schelle, "World's Top 10 Rivers at Risk" (World Wildlife Fund International, Gland, Switzerland, 2007), p. 28; http://assets.panda.org/downloads/worldstop10riversatriskfinalMarch13.pdf.

5. Maude Barlow and Tony Clarke, *Blue Gold: The Fight to Stop the Corporate Theft of the World's Water* (New York: The New Press, 2002), p. 7.

6. Heather Cooley, P. Gleick, and Gary Wolff, *Desalination, with a Grain of Salt: A California Perspective* (Oakland, Calif.: Pacific Institute, 2006), p. 21.

7. Emily Yuhas and Tom Daniels, "The US Freshwater Supply Shortage: Experiences with Desalination as Part of the Solution," *Journal of Environmental Planning and Management* 49 (2006): 575.

8. Ibid., pp. 578–80.

9. Cooley et al., *Desalination, with a Grain of Salt*, p. 54.

10. Sue McClurg, "Urban Growth and Water: States Seek New Supplies," *River Report* (Colorado River Project), Fall 2007, pp. 1, 4–11.

11. Cooley et al., *Desalination, with a Grain of Salt*, p. 28, box 4.

12. Yuhas and Daniels, "The US Freshwater Supply Shortage," p. 573.

13. Ibid., p. 578.

14. Cooley et al., *Desalination, with a Grain of Salt,* p. 24.

15. Craig Pittman, "Water Plant Delays Run Patience Dry," *St. Petersburg Times,* 18 December 2006.

16. Ibid.

17. Neil Johnson, "Agency Settles Desalination Plant Lawsuit," *Tampa Tribune,* 17 April 2007.

18. Cooley et al., *Desalination, with a Grain of Salt,* p. 24.

19. Yuhas and Daniels, "The US Freshwater Supply Shortage."

20. Ibid., p. 576.

21. Henry Fountain, "The Science of Rain-Making Is Still Patchy," *New York Times,* 19 October 2003.

22. Jeff Hull, "Idea Lab: The Modern Rain Dance," *New York Times,* 2 July 2006.

23. Theodore W. Fuller, *San Diego Originals: Profiles of the Movers and Shakers of California's First Community* (Pleasant Hill, Calif.: California Profiles Publications, 1987), pp. 51–53.

24. National Research Council, *Critical Issues in Weather Modification Research* (Washington, D.C.: National Research Council, 2003), p. 3.

25. Launce Rake, "New Look at Old Idea Could Be a Rainmaker," *Las Vegas Sun,* 6 July 2006; D. A. Griffith and M. E. Solak, "Upper Colorado Basin: Technical Feasibility," *Southwest Hydrology,* March/April 2007.

26. Associated Press, "Colorado River Basin States Look to Regional Cloud-Seeding," 13 March 2006.

27. Quoted in Kate Ravilious, "Kicking Up a Storm with the Cloud Seeders," *New Scientist,* 16 April 2005.

28. Quoted in Jerd Smith, "The West's Water Future May Float on Cloud Plan," *Rocky Mountain News,* 17 April 2006.

29. Bureau of Reclamation, "Weather Modification at the Bureau of Reclamation" (2006), http://www.usbr.gov/pmts/rivers/WxMod.html.

30. Tim Barnett, Robert Malone, William Pennell, Detlet Stammer, Bert Semtner, and Warren Washington, "The Effects of Climate Change on Water Resources in the West: Introduction and Overview," *Climatic Change* 62 (2004).

31. P. W. Mote, "Trends in Snow Water Equivalent in the Pacific Northwest and Their Climatic Causes," *Geophysical Research Letters* 30 (2003): 1601.

32. Barnett et al., "Effects of Climate Change on Water Resources in the West," p. 7.

33. Quoted in Robert F. Service, "As the West Goes Dry," *Science* 303 (2004): 1126.

34. Marc Reisner, *Cadillac Desert: The American West and Its Disappearing Water* (New York: Penguin Books, 1987), p. 507.

35. Ibid., p. 512.

36. A. F. Pillsbury, "The Salinity of Rivers," *Scientific American* 245 (1981).

37. Quoted in Reisner, *Cadillac Desert,* p. 510.

38. P. H. Gleick, "Global Freshwater Resources: Soft-Path Solutions for the 21st Century," *Science* 302 (2003): 1524–28.

CHAPTER SIXTEEN

1. National Research Council, *Colorado River Ecology and Dam Management: Proceedings of a Symposium, May 24–25, 1990, Santa Fe, New Mexico* (Washington, D.C.: National Academy Press, 1991), p. 63.

2. Ibid., p. 57.

3. R. L. Ferrari, *1986 Lake Powell Survey* (Denver, Colo.: Bureau of Reclamation, 1988).

4. Quoted in Marc Reisner, *Cadillac Desert: The American West and Its Disappearing Water* (New York: Penguin Books, 1987), p. 492.

5. A report from Living Rivers titled *The One-Dam Solution* uses the same kind of analysis and comes to a similar conclusion.

6. Lynn Merrill, "The Arnold Schwartzenegger of Trucks," *Grading and Excavation Contractor,* July/August 2002.

7. Aldo Leopold, Susan Flader, and J. Baird Callicott, *The River of the Mother of God and Other Essays* (Madison: University of Wisconsin Press, 1991), p. 179.

8. Matilija Dam Ecosystem Restoration Project (2006), http://www.matilijadam.org; Paul Jenkin, "The Matilija Coalition" (2002), http://pages.sbcglobal.net/pjenkin/matilija/index.htm.

9. Matilija Dam Ecosystem Restoration Project, *Restoration Report,* http://www.matilijadam.org, pp. 1–2.

10. Ibid.

11. Quoted in Richard J. Ingebretsen, "Sedimental Journey: A Grim Prospect for Lake Powell" (1998), http://www.glencanyon.org.

12. Reisner, *Cadillac Desert,* p. 493.

13. A. F. Pillsbury, "The Salinity of Rivers," *Scientific American* 245 (1981).

14. Bureau of Reclamation, "Law of the River" (2006), http://www.usbr.gov/lc/region/pao/lawofrvr.html.

15. Ibid.

16. Bureau of Reclamation, Department of Agriculture, and Department of the Interior, *Quality of Water, Colorado River Basin: Progress Report,* No. 22 (Washington, D.C.: Department of the Interior, 2005).

17. Ibid., p. 26.

18. L. L. Nash and P. H. Gleick, *The Colorado River Basin and Climate Change: The Sensitivity of Streamflow and Water Supply to Variations in Temperature and Precipitation* (Washington, D.C.: U.S. Environmental Protection Agency, 1993), p. 68.

CHAPTER SEVENTEEN

1. *Colorado River Storage Project,* 43 U.S. Code 620.

2. The figure of 8.23 MAF is the sum of 7.5 MAF to meet the Colorado River Compact plus 0.75 MAF to meet the upper basin's half of the debt to Mexico, minus 20,000 acre-feet to take into account the contribution of Arizona's tributaries to the Colorado River.

3. *Colorado River Basin Project* (italics added).

4. D. H. Getches, "Colorado River Governance: Sharing Federal Authority as an Incentive to Create a New Institution," *University of Colorado Law Review,* Summer 1997.

5. Ibid.

6. National Research Council, *River Resource Management in the Grand Canyon* (Washington, D.C.: National Academy Press, 1996), p. 16.

7. Getches, "Colorado River Governance," p. 9.

8. Quoted in Byron Hayes, "Introducing Secretary of the Interior James Watt," *The Waiting List* 3, no. 2 (1999): 10.

9. Quotes in George Sibley, "Glen Canyon: Using a Dam to Heal a River," *High Country News,* 22 July 1996.

10. National Research Council, *River Resource Management in the Grand Canyon*, p. 76.

11. Getches, "Colorado River Governance."

12. *Grand Canyon Protection Act,* Public Law 93–620, 88 Stat. 2089.

13. National Research Council, *Colorado River Ecology and Dam Management: Proceedings of a Symposium, May 24–25, 1990, Santa Fe, New Mexico* (Washington, D.C.: National Academies Press, 1991), p. 124.

14. Steven Gloss, Jeffrey E. Lovich, and T. S. Melis, "The State of the Colorado River Ecosystem in the Grand Canyon: U.S. Geological Survey Circular 1282" (2005), p. 33.

15. Ibid., p. 18.

16. Ibid., p. 22.

17. D. J. Topping, D. M. Rubin, J. C. Schmidt, J. E. Hazel, S. A. Wright, T. S. Melis, and M. Kaplinski, "Comparison of Sediment-Transport and Bar-Response Results from the 1996 and 2004 Controlled-Flood Experiments on the Colorado River in Grand Canyon" (abstract H52A-06, supplement to American Geophysical Union meeting, San Francisco, Fall 2005).

18. Gloss, Lovich, and Melis, "The State of the Colorado River Ecosystem in the Grand Canyon."

19. Ibid., p. 37.

20. Joe Baird, "Groups to File Suit over Decline of Fish below Grand Canyon," *Salt Lake Tribune,* 9 November 2005.

21. Quotes in ibid.

22. Gloss, Lovich, and Melis, "The State of the Colorado River Ecosystem in the Grand Canyon," p. 28.

23. T. J., Randle, J. K. Lyons, R. J. Christensen, and R. D. Stephen, *Colorado River Ecosystem Sediment Augmentation Appraisal Engineering Report*, ed. Bureau of Reclamation (Denver, Colo.: Bureau of Reclamation, 2007).

24. Joe Baird, "Dam Still Injuring Grand Canyon," *Salt Lake Tribune*, 27 October 2005.

25. U.S. Geological Survey, *Grand Canyon Humpback Chub Population Improving*, Fact Sheet 2007-3113 (December 2007).

26. Philip L. Fradkin, *A River No More: The Colorado River and the West* (Berkeley: University of California Press, 1996).

27. Aldo Leopold, *A Sand County Almanac: With Essays on Conservation from Round River* (New York: Ballantine Books, 1970), p. 150.

28. Ibid., p. 151 (Leopold's italics).

29. Jeffrey P. Cohn, "Colorado River Delta," *Bioscience* 54 (2004): 388.

30. D. F. Luecke and J. Pitt, *A Delta Once More: Restoring Riparian and Wetland Habitat in the Colorado River Delta* (Boulder, Colo.: Environmental Defense Fund, 1999).

31. Charles Bergman, *Red Delta: Fighting for Life at the End of the Colorado River* (Golden, Colo.: Fulcrum Publishers, 2002), p. 50.

32. Mitch Tobin, "Delta Beast Rears Its Head," *High Country News*, 15 September 2003.

33. Ibid.

CHAPTER EIGHTEEN

1. The apt reversal of phrase used for this chapter title, from "Law of the River" to "River of Law," comes from Charles Bergman, *Red Delta: Fighting for Life at the End of the Colorado River* (Golden, Colo.: Fulcrum Publishers, 2002).

2. Bureau of Reclamation, "Law of the River" (2006), http://www.usbr.gov/lc/region/pao/lawofrvr.html.

3. R. Eric Kuhn, general manager of the Colorado River Water Conservation District, has also speculated about the future of the Colorado River. See Kuhn, "Future Scenarios for the Colorado River" (paper presented at the American Bar Association, San Diego, February 24–25, 2005).

4. The Sonoran Institute, in Tucson, describes four scenarios for the future of the Colorado River; see "Ecosystem Changes and Water Policy Choices: Four Scenarios for the Lower Colorado River Basin to 2050" (2007), http://sonoran.org. The "dry future" scenario assumes that climate change reduces the flow of the river by 40 percent by 2050.

5. Christopher Smart, "Glen Canyon: 50 Years of Controversy," *Salt Lake Tribune*, 30 May 2007.

6. Tom Ross, "Downriver Deals," *Steamboat Pilot & Today*, 20 August 2006.

7. Jeffrey P. Cohn, "Colorado River Delta," *Bioscience* 54 (2004).

8. Launce Rake, "Agency's Water Plans Pegged at $7.6 Billion," *Las Vegas Sun,* 24 May 2005.

9. Bureau of Reclamation, *Final Environmental Impact Statement, Colorado River Interim Guidelines for Lower Basin Shortages and Coordinated Operations for Lake Powell and Lake Mead* (Washington, D.C.: Department of the Interior, 2007), vol. 4, pp. 4–9.

10. Southern Nevada Water Authority, "Virgin and Muddy Rivers Surface Water Development Project" (2006), http://www.snwa.com.

11. Matt Jenkins, "'Tributary' Issue Could Force a Seven-State Showdown," *High Country News,* 19 September 2005.

12. Peter Sartucci, "'Big Daddy Drought' Will Be a Complicated Matter," *High Country News,* 4 September 2006, letter to the editor and editor's reply.

13. Colorado River Commission, "Colorado River Compact" (1922), http://www.colorado.edu/resources/colorado_river/law_policy.html.

14. Bureau of Reclamation, *The Colorado River: A Natural Menace Becomes a National Resource* (Washington, D.C.: Department of Interior, 1946).

15. Kuhn, "Future Scenarios for the Colorado River."

16. City of Phoenix, "Water Resources Plan: 2005 Update" (Water Services Department, 2005), p. 63.

17. U.S. Geological Survey, "Ground-Water Depletion" (2005), http://ga.water.usgs.gov/edu/gwdepletion.html.

18. Matt Jenkins, "The Perpetual Growth Machine," *High Country News,* 12 June 2006.

CHAPTER NINETEEN

1. *Arizona Daily Star,* "Developers Plan Massive Water Park in Arizona Desert" (2007), http://www.azstarnet.com/news/212479.

2. Environmental Working Group, *California Water Subsidies* (2007), www.ewg.org.

3. *Economist Magazine,* "The Big Dry," 28 April 2007.

4. California Department of Water Resources, *California Water Plan: A Framework for Action,* Bulletin 160-05 (December 2005).

5. P. H. Gleick, *Dam It, Not Now. Testimony before the California Senate Committee on Natural Resources and Water* (2006), available at www.pacinsti.org/press_center.

6. *Congressional Record,* "Remarks of Francis Newlands," 13 June 1902, p. 6734.

7. Bernard DeVoto, "The West against Itself," *Harpers,* January 1947.

INDEX

dams (continued)
    43–44, 47, 54; failures of, 5, 47, 72;
    inevitable future of, 160, 206–9; in
    national parks/monuments, 101, 103–4,
    139–40; raising height of existing, 198
Davis, Arthur Powell, 45, 49, 56, 60, 64, 74
dead pool: defined, 18; for Lake Mead, 184;
    for Lake Powell, 8, 19, 184, 185, 188
A Delta Once More (Environmental
    Defense), 223
dendrochronology: and BuRec's forecasts
    of future, 181–83; Colorado River flow
    as revealed by, 33, 166–68; invention
    of, 165–66
desalination, 189–95; appeal of, 189;
    encouraged by 2007 agreement, 187;
    largest U.S. plants for, 192–94,
    223–24; problems with, 190–91, 194
Desert Land Act, 40, 47, 63
DeVoto, Bernard, 106–7, 108, 146, 246
Dexheimer, W.A., 118
Dinosaur National Monument: creation
    of, 91; Echo Park in, 97; lack of famil-
    iarity with, 92, 98, 105; "low Glen" vs.
    "high Glen" options for dams in,
    116–18; proposed downgrade to
    national recreation area, 100–101;
    Sierra Club efforts to publicize, 114,
    119–20. See also Echo Park dam
Doe, Phil, 154
Dominy, Floyd: biography of, 143, 144;
    and dams in Grand Canyon, 135, 141,
    142; disliked Brower, 140; on draining
    Lake Powell, 146–47; on environmen-
    tal protection legislation, 112, 114;
    Lake Powell: Jewel of the Colorado, 136;
    named Lake Powell, 20; photo of, 130;
    and protection of Rainbow Bridge,
    129–31; raft trip with Brower and
    McPhee, 144, 146; on reason for
    building Glen Canyon Dam, 157; on
    recreation at Lake Powell, 159; on sed-
    iment in Lake Powell, 209
Douglas, Paul, 148–51, 214, 245
Douglass, Andrew E., 163, 165–66
Dreyfus, Daniel, 153
drought: early 21st-century, 18–19, 169;
    Glen Canyon Dam as protection

from, 157–58; late 1880s, 45; predic-
    tions of, in Southwest, 22, 180;
    response to, with Law of the River,
    19, 157–58; revealed by den-
    drochronology, 166, 168; and vanish-
    ing Native American civilizations, 33,
    166. See also global warming
Drury, Newton, 100, 102, 103, 107, 109
Dyrenforth, R.G., 195

Echeverría, Luis, 210
Echo Park, 97, 98
Echo Park dam: estimates of evaporation
    from, 102, 103, 104, 115–18, 127; leg-
    islative debate over, 100–105, 114–20,
    126–27; national publicity on, 8,
    106–7; proposed as CRSP cash-regis-
    ter dam, 96–98; removed from CRSP,
    121, 126; and Sierra Club, 98, 114, 115,
    116–17, 118, 119, 120, 127
Egypt, irrigation-based society of, 29–32
Eisenhower, Dwight D., 96, 118, 126
Ely, Northcutt, 82, 120
Empire State Building, 79
Endangered Species Act, 145, 212, 218, 220,
    224–25
Environmental Defense (formerly
    Environmental Defense Fund), 109,
    215, 223
environmental impact statement (EIS): on
    Colorado River system with pro-
    longed drought, 182–83; environmen-
    tal assessment as alternative to, 216;
    Glen Canyon Dam built without, 112,
    215; for Glen Canyon operations, 217;
    on humpback chub, 220
Escalante National Monument, 92
Espy, James Pollard, 195
evaporation: as concern of LaRue, 65; with
    Echo Park dam, 102, 103, 104, 115–18,
    127; estimated annual, from Colorado
    River, 169; vs. regulation gain, 155–57
evapotranspiration, defined, 179–80

Fall, Albert, 64
farmers: and early Reclamation Service
    efforts, 52–57; Imperial Valley, recla-
    mation law requirements ignored by,

81–83; limits on public land for, 40, 42, 49, 52–53, 81–82; poor, as concern of Powell, 40–41, 42, 53, 80. *See also* agriculture

federal reclamation program. *See* Bureau of Reclamation; Reclamation Service

fish: improving conditions for, 214; native, 212–13, 217–18, 220, 221; rainbow trout, 213

Flaming Gorge Dam, 126, 155–56

floods: of 1980s, and Colorado River Delta, 223; "controlled," from Glen Canyon Dam releases, 14–15, 219, 221; from dam breaks, 5, 47, 72; flash, as source of silt, 203; forming Salton Sea, 63; and South Dakota cloud seeding, 197

Florida, Tampa Bay desalting plant, 193–94

Fradkin, Philip, 222

Gabrielson, Ira, 103

Gamble, Tom, 5

Garfield, James R., 56

General Land Office, 40

Geological Survey. *See* U.S. Geological Survey

Gila River, 33, 61, 70, 71, 192

Glacier National Park, 103

Gleick, Peter, 200

Glen Canyon: beauty of, 20, 67, 128, 135–36, 137; LaRue's recommendation for dam at, 65, 67, 70; proposed national monument to include, 92; revealed by falling level of Lake Powell, 20; as site benefiting upper basin states, 89

Glen Canyon Dam: benefits of, 155–59; BuRec report preceding, 90, 93, 94, 95, 96; construction of, 79, 80, 122, 123, 124, 125; "controlled flood" releases from, 14–15, 219, 221; cost of, 149; CRSP legislation authorizing, 96–98, 126; future of, 160, 209; Hoover Dam as model for, 72, 78; impact on Colorado River in Grand Canyon, 212–13, 216; impact on water reaching Mexico, 210, 223; LaRue's proposal for, 65, 67, 70, 135; "low Glen" vs. "high Glen" option for,

116–18; power generation as purpose of, 6–7, 132, 158, 213–14, 215; safety of site for, 123–26; and Sierra Club, 110, 127–28, 135; spillway design of, 10, 11–13, 16, 78; spring 1983 crisis at, 3–17; studies of environmental impact of, 112, 215–17. *See also* Powell, Lake

Glen Canyon Environmental Studies, 216, 218

Glen Canyon Institute, 146, 220

Glen Canyon National Recreation Area, 20, 159, 217

Glenn, Ed, 224

global warming: calculations/forecasts ignoring, 19, 31, 181–83; cause of, 171–74, 182, 184; dams to combat, 23, 244–45; effect on snowmelt/runoff predictions, 21–22, 177–78; evident in Southwest, 22–23, 241; future of Colorado River system with, 22–23, 176–78, 231–40, 270n4; impossibility of stopping, 175–76; scientific evidence of, 171; understatement of dangers of, 174–75; White House on, 184. *See also* drought

Goldwater, Barry, 98

Gore, Al, 184

Grand Canyon: "controlled floods" in, 14–15, 219, 221; fish in Colorado River in, 212–13, 217–18, 220, 221; and fluctuating water releases from Glen Canyon Dam, 213, 216; proposed dams in, 103, 109–11, 137–41; publicity about damming, 138–39; sandbars in, 213, 218–19, 221

Grand Canyon Act, 138

Grand Canyon National Park, 109–10, 140, 144, 217

Grand Canyon Protection Act, 217, 220

Grand Coulee Dam, 74, 88

Grand Staircase–Escalante National Monument, 92–93

Grant, Ulysses S., 40

Grant, Ulysses S. III, 104, 116, 150

Gray, Stephen, 167–68

Green River: and Echo Park dam, 97, 100, 101, 119; Flaming Gorge Dam on, 126, 155–56; rafters on, 91–92, 114

of, 53, 56; failures of early efforts of, 53–57; multiple early projects of, 51–52; Salt River project of, 52–53. *See also* Bureau of Reclamation

recreation: increased interest in, 101; Lake Powell's use for, 19–20, 158–59

regulation: as benefit of reclamation projects, 155; vs. evaporation, 155–57

Reisner, Marc, 138, 153, 154, 199, 204–5

river-basin accounting, 94–95, 148

Robinson, Michael, 53–54

Rockwood, C.R., 62, 63

Roosevelt, Franklin Delano, 78, 91, 92

Roosevelt, Theodore, 48, 52–53, 91

Roosevelt Dam, 52, 58

runoff: defined, 179; forecasts of, 7, 180–83, 185; and increased evapotranspiration, 179–80. *See also* snowmelt; snowpack

Rusho, W.L., 11, 13

Rutledge, Burl, 77

Safarin, Robert, 197

salt: in Colorado River water reaching Mexico, 209–11; as enemy of irrigation societies, 28, 31. *See also* desalination

Salton Sea, 63

Salton Trough, 202

Salt River project, 52–53, 245

Santa Barbara, CA, desalting plant, 192

Saylor, John, 131

Schaefer, Vincent, 196

Seaton, Fred, 157

Second Hoover Commission, 151–52

sedimentation. *See* silt

sediment gauges, 203

Shasta Dam, 80, 198

Shea, Charlie, 73

Sierra Club: books on Glen Canyon published by, 91, 135–36; and Bridge Canyon dam, 135, 140; changed mission and membership requirements of, 111; Dinosaur National Monument publicized by, 114, 119–20; and Echo Park dam, 98, 114, 115, 116–17, 118, 119, 120, 127; and Hetch Hetchy, 109; and humpback chub in Grand Canyon, 220; and proposed dams in Grand Canyon, 98, 109–10, 138–39; and

protection of Rainbow Bridge, 133. *See also* Brower, David

silt, 202–8; carried by Colorado River, 61–62, 63, 77, 202–4; and Colorado River Delta, 222; counteracting build-up of, 204–6; deposited in Lake Powell, 203–4, 205–6; as enemy of irrigation societies, 28, 31, 63; in Grand Canyon, 212, 219; inevitable future of dams due to, 160, 206–9

Sites Reservoir, 244

Six Companies: Hoover Dam construction by, 74–80; member companies, 73–74

Smith, Joseph, 34

Smythe, William, 62

snowmelt: earlier, with global warming, 177–78; and May 1983 crisis at Glen Canyon Dam, 6, 10, 17; reduced, with higher temperatures, 21–22, 169, 179, 245; surge of water volume with, 213, 244–45. *See also* runoff

snowpack: decline in, 198, 245; effect of reduced, 174, 177; vanishing, in spring 2006, 20, 21. *See also* precipitation

Sonoran Institute, 225, 270n4

South Dakota, cloud seeding, 197

South Fork Dam, 47

Split Mountain, 101, 103, 104, 116, 257n6

Stanton, Robert, 45

The State of the Colorado River Ecosystem in the Grand Canyon (Geological Survey), 219–21

Stegner, Wallace: on aridity of West, 42; on beauty of Lake Powell vs. Glen Canyon, 136–37; on Glen Canyon vs. Grand Canyon, 20; on Homestead Act, 40; on Powell, 38, 46–47; *This Is Dinosaur: Echo Park Country and Its Magic Rivers*, 120, 128

St. Francis Dam, failure of, 5, 72

Stockton, Charles, 166–67

Straus, Michael W., 90–91, 96–97, 102, 107

suckers, razorback, 218

Sumerians, 27–28, 29

Swampland Act, 40, 47

Swing, Phil, 66

Swing-Johnson bill, 71, 81

Widtsoe, John, 38, 56, 59, 67
Wilbur, Ray Lyman, 73, 82
Wilderness Act, 126
wildfires, 177
Wiley, Lem, 129
Wilson, Woodrow, 91
Wolf, T. J., 5, 14–15
Woodhouse, Connie, 167–68
Worster, Donald, 29, 38, 40, 45, 49, 55
Wright Act, 47
Wyoming: cloud seeding, 196; early state-funded irrigation program in, 48; and

Upper Colorado River Basin Compact, 96
Wyoming v. Colorado, 66

Yampa River, 97, 100, 101. *See also* Echo Park dam
Yellowtail Dam, 10–11
Young, Brigham, 34
Yuma, AZ, desalting plant, 192, 223–24

Zahniser, Howard, 126

Text:	11.25 × 13.5 Adobe Garamond
Display:	Adobe Garamond
Compositor:	International Typesetting and Composition
Indexer:	Jean Mann
Cartographer:	Bill Nelson
Printer and binder:	Sheridan Books, Inc.